Reservoir Management

Reservoir Management

A Practical Guide

Steve Cannon

Principal Consultant
Steve Cannon Geoscience
UK

WILEY Blackwell

This edition first published 2021
© 2021 John Wiley and Sons Ltd

Registered Offices
John Wiley & Sons, Inc., 111 River Street, Hoboken, NJ 07030, USA
John Wiley & Sons Ltd, The Atrium, Southern Gate, Chichester, West Sussex, PO19 8SQ, UK

Editorial Office
9600 Garsington Road, Oxford, OX4 2DQ, UK

For details of our global editorial offices, customer services, and more information about Wiley products visit us at www.wiley.com.

Wiley also publishes its books in a variety of electronic formats and by print-on-demand. Some content that appears in standard print versions of this book may not be available in other formats.

Library of Congress Cataloging-in-Publication Data

Name: Cannon, Steve, 1955– author.
Title: Reservoir management : a practical guide / Steve Cannon.
Description: Hoboken, NJ : Wiley-Blackwell, 2021. | Includes
    bibliographical references and index.
Identifiers: LCCN 2020036606 (print) | LCCN 2020036607 (ebook) | ISBN
    9781119619369 (hardback) | ISBN 9781119619413 (adobe pdf) | ISBN
    9781119619437 (epub)
Subjects: LCSH: Gas reservoirs.
Classification: LCC TN880 .C36 2021 (print) | LCC TN880 (ebook) | DDC
    622/.3385–dc23
LC record available at https://lccn.loc.gov/2020036606
LC ebook record available at https://lccn.loc.gov/2020036607

Cover Design: Wiley
Cover Image: © Steve Cannon

Set in 9.5/12.5pt STIXTwoText by SPi Global, Pondicherry, India
Printed and bound by CPI Group (UK) Ltd, Croydon, CR0 4YY

10 9 8 7 6 5 4 3 2 1

To my parents, Bill and Biddy Cannon, to whom I owe so much and more!

Contents

Preface

I am a geologist by profession, a petrophysicist by inclination, and a reservoir modeler by design, but I have always wanted to be a reservoir engineer; they get paid more as a rule and usually end up as project managers! In 2018, while on holiday, I received an email asking me to deliver a course in India on reservoir management. My initial response was that I did not have the necessary skills and that the course organizer should look for an experienced reservoir engineer. As I relaxed in the sun I started to think about a new project during the latest oilfield downturn, and I had the idea for a book on reservoir management, but one that would follow the same principles as my guides to petrophysics and reservoir modeling: a geologist's view on the topic. Fortunately, Wiley-Blackwell were keen for me to submit a proposal, which I duly did, but with the proviso that I might find out that I was out of my depth!

Researching which publications were available, I realized that *Integrated Petroleum Reservoir Management: A Team Approach* by Satter and Thakur (1994) was around 25 years old and that nothing newer was available; there were innumerable peer reviewed articles in the journals covering most of the aspects of reservoir management and a number of commercial and academic courses but no single book: here was my opportunity! This book is not a deeply academic tome but rather the description of a process enlivened by a number of stories and case studies that I have come across over 40 years of experience in the oil patch. At one time I had a working title of 'Tales of the Unexpected' but I realized Roald Dahl had beaten me to it!

A word of caution; my formal training in the oil industry was a series of courses given by my employer through the 1980s, Shell (UK) Exploration & Production Ltd. (Shell Expro). I studied production geology, petrophysics, and reservoir engineering (PE101) and was taught that oil was always red and gas was green in maps, well sections, and cross plots. The rest of the industry uses these colors the other way around, but I remain true to my training and the company that pays me a small monthly pension. 'There is a right way to do something, a wrong way and the Shell way' is as valid today as ever!

I am deeply grateful to all the reservoir engineers I have worked with over the years who have had the patience to explain how an oil or gas field should be managed: gently, with due reverence for its uncertain and sometimes unknowing response to your ministrations and demands, just like your life partner in fact! In particular, I would like to thank Steve Griffith, Alun Griffiths, Steve Flew, Pat Neve, Andrew Evans, Dave Ponting, Steve Furnival, Dan O'Meara, Jez Christiansen, Christian Masini, and Neil Ementon.

Steve Cannon
2021

List of Abbreviations

AAPG	American Association of Petroleum Geologists
AHD	Along hole depth
AI	Acoustic impedance
API	American Petroleum Institute
AVO	Amplitude versus offset
BBO	Billion barrels of oil
BHA	Bottomhole assembly
BHFP	Bottomhole flowing pressure
BHT	Bottomhole temperature
BOPD	Barrels of oil per day
BSW	Bulk solids and waste
CAPEX	Capital expenditure
CGR	Condensate–gas ratio
DCD	Downhole control devices
DFE	Drill floor elevation
DHS	Downhole sensors
DHVT	Downhole vibration tool
EMV	Expected monetary value
EOR	Enhanced oil recovery
EOS	Equation of state
ERD	Extended reach drilling
ESP	Electric submersible pumps
EUR	Estimated ultimate recovery
EWT	Extended well tests
FA	Forties Alpha

FB	Forties Bravo
FC	Forties Charlie
FDP	Field development plan
FEED	Front end engineering design
FFM	Full-field model
FID	Final investment decision
FIPNUM	Fluid in-place region numbers
FLAGS	Far-north Liquids and Associated Gas System (North Sea gas pipeline)
FMT	Formation measurement tool (wireline formation tester)
FPS	Forties pipeline system
FPSO	Floating production, storage, and offloading
FUKA	Frigg UK Association pipeline
FVF	Formation volume factor
FWL	Free water level
GIIP	Gas initially in-place
GOC	Gas–oil contact
GOR	Gas–oil ratio
GOSP	Gas–oil separation plant
GR	Gamma ray
GRV	Gross rock volume
GWC	Gas–water contact
HIIP	Hydrocarbons initially in-place
ICD	Inflow control devices
ICV	Interval control valves
IMPES	Implicit pressure explicit saturation
IOR	Improved oil recovery
IRR	Internal rate of return
LPG	Liquid petroleum gas
LWD	Logging-while-drilling
MDT	Modular dynamic tester
MEG	Mono-ethylene glycol
MEOR	Microbial enhanced oil recovery
MMbbls	Millions of barrels
MMboe	Million barrels of oil equivalent
MMscf/d	Million standard cubic feet per day
MNCF	Maximum negative cash flow
MPS	Multipoint statistical methods

MWD	Measurement-while-drilling
NMR	Nuclear magnetic resonance
NNM	Not normally manned
NPV	Net present value
NRI	Net revenue interest
NRV	Net rock volume
NTG	Net-to-gross
OBC	Ocean bottom cable
OBN	Ocean bottom node
OPEX	Operating expenditure
OWC	Oil–water contact
PDG	Permanent downhole gauges
PdVSA	Petroleos de Venezuela
PEBI	Perpendicular–bisectional grid system
PI	Production index
P/I	Profit-to-investment ratio
PLT	Production logging tool
PRMS	Petroleum Resources Management System
PSDM	Pre-stack depth migration
PSA	Production sharing agreement
PVT	Pressure–volume–temperature
QC	Quality control
REV	Representative elementary volume
RFT	Repeat formation tester
SEG	Society of Exploration Geophysicists
SIS	Sequential indicator simulation
SPE	Society of Petroleum Engineers
SPEE	Society of Petroleum Evaluation Engineers
SSSV	Subsurface safety valve
STOIIP	Stock tank oil initially in-place
TCFG	Trillion cubic feet of natural gas
TDT	Thermal decay time
TGSim	Truncated Gaussian simulation algorithm
THP	Tubing head pressures
TPM	Total property modeling
TST	True stratigraphic thickness
TTRD	Through tubing rotary drilling
TVD	True vertical depth

TVDSS   True vertical depth subsea
UKCS   United Kingdom continental shelf
UTM   Universal Transverse Mercator coordinate system
VFP   Vertical flow performance
VOI   Value of information
WAG   Water alternating gas
WFT   Wireline formation tester
WGR   Water–gas ratio
WOR   Water–oil ratio
WPC   World Petroleum Council

1

Introduction

Reservoir management is fundamental to the efficient and responsible means of extracting hydrocarbons, and maximizing the economic benefit to the operator, license holders, and central government. All stakeholders have a social responsibility to protect the local population and environment. The process of managing an oil or gas reservoir begins after discovery and continues through appraisal, development, production, and abandonment (Figure 1.1); there is cost associated with each phase and a series of decision gates should be in place to ensure that an economic benefit exists before progress is made. To correctly establish potential value at each stage it is necessary to acquire and analyze data from the subsurface, the planned surface facilities, and the contractual obligations to the end-user of the hydrocarbons produced. This is especially true of any improved recovery methods proposed or plans to extend field life. To achieve all the above requires a multiskilled team of professionals working together with a clear set of objectives and associated rewards. The team's make-up will change over time as different skills are required, as well as the management of the team, with geoscientists, engineers, and commercial analysts needed to address the issues as they arise.

This book is designed as a guide for nonspecialists involved in the process of reservoir management, which is often treated as a task for reservoir engineers alone. It is a task for all the disciplines involved in turning an exploration success into a commercial asset. Most explorers earn their bonus based on the initial estimates of in-place hydrocarbons, regardless

Reservoir Management: A Practical Guide, First Edition. Steve Cannon.
© 2021 John Wiley & Sons Ltd. Published 2021 by John Wiley & Sons Ltd.

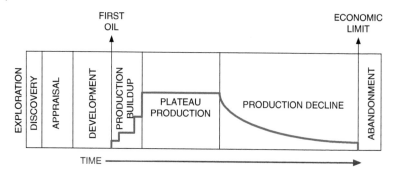

Figure 1.1 Oilfield life cycle from discovery to abandonment with a typical primary production profile in red. The period from discovery to first oil may be short or long depending on economic conditions and infrastructure limitations.

of the ultimate cost of production; the explorers have usually moved on to a new basin before the first oil or gas is produced!

This chapter will look at the basics of reservoir management introducing the main terms and jargon, while subsequent chapters will go into more detail.

- Chapter 2 reviews the life cycle of an oil or gas field after discovery, looks at field development plans, monitoring and data acquisition requirements, and discusses these issues in the light of a number of case studies.
- Chapter 3 looks at the static and dynamic reservoir description around which the initial plans are built.
- Chapter 4 reviews the construction of the integrated reservoir model.
- Chapter 5 addresses reservoir performance and production forecasting, reviewing the dynamic estimation and uncertainty in future resources and reserves.
- Chapter 6 discusses some of the ways to improve or enhance hydrocarbon recovery with examples used to describe the methods.
- Chapter 7 focuses on the economic aspects of a successful field development and how active reservoir management through the field life can improve the returns on investment.
- Chapter 8 considers the way in which the reservoir management plan evolves with time from project sanction to abandonment. We will look at some of these issues in the real world of field development and secondary recovery projects through a series of case studies.

Throughout the book there are relevant examples from real reservoir management projects and field developments across the major hydrocarbon basins of the world. I have also included an Appendix that covers the basics of dynamic reservoir simulation, which is the main tool used by those involved in reservoir management studies.

1.1 The Basics

The main objectives of reservoir management may be summarized as follows:

- Maximizing the ultimate recovery of reserves from an oil or gas field
- Reducing the commercial risk associated with development plans
- Minimizing operating expenditure (OPEX) and capital expenditure (CAPEX)
- Increasing hydrocarbon production from wells
- Increasing the value of reserves through time (net present value – NPV)

To maximize the economic recovery of hydrocarbons requires the identification and characterization of all potential reservoirs in a field so that the optimum development plan can be proposed. This requires a reservoir management plan designed on the basis of location and size of the field, the geological complexity of the reservoir, the type and distribution of the reservoir rock and fluids, the drive mechanism, regulatory controls and contractual limitations, and economics. Different management plans are required for onshore and offshore locations, and also for gas-filled 'tanks of sand' and poorly connected oil reservoirs, especially when natural depletion results in economically low recovery.

To achieve these objectives requires the integration of static and dynamic models of the reservoir together with gross uncertainties associated with a complex natural system, as well as models of surface facilities designed to optimize production from the field, economic models of OPEX, CAPEX, and price fluctuations throughout field life. The uncertainty associated with each of these inputs leads to a range of potential rewards; determining the relative value of these outcomes is the task of the whole team at the time of evaluation and prior to any investment being made.

The life cycle of an oil or gas field begins after discovery with a clear appraisal plan, to delineate the field, upon which the development program is designed, costed, and approved. Inevitably, surprises will occur during development drilling that require a change to the plan, but these should have been considered and included in the budget. An oil or gas field only begins to make money once production has started: most oil companies expect a return on investment within three to five years of this date. Thereafter, the thoughts move toward improving overall returns through innovative secondary recovery techniques or increasing ultimate recovery using more esoteric tertiary methods. Effective reservoir management should always be proactive, anticipating decline, and investing to maintain production and improve recovery through to abandonment (Figure 1.2). Reacting to decline after it has started may limit the solutions available and ultimately cost more. Both of these phases are often invoked to delay the ultimate stage of the life cycle, cessation of production and then abandonment, the costs of which should also have been included in the original development plan.

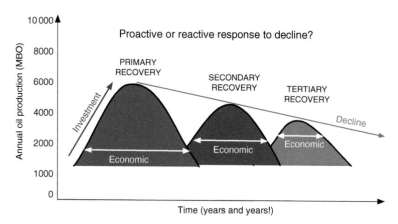

Figure 1.2 Effective reservoir management should always be proactive, anticipating decline, and investing to maintain production and improve recovery through to abandonment. Reacting to decline after it has started may limit the solutions available and ultimately cost more.

1.2   Field Appraisal

As soon as an oil, condensate, or gas field is discovered there is internal and external pressure to monetize what will have been a significant investment in the exploration phase. The appraisal plan is drawn up to try to reduce the major uncertainties in the reservoir description and thus increase confidence in the value of the potential project. A lot will depend on the data collected during the exploration phase: was the reservoir cored and logged to determine rock properties? was a well test carried out and fluid samples collected? was the well suspended or abandoned? Many operators never core or test an exploration well; they delay data gathering to the next phase. In an onshore location this may be understandable, but offshore, where the well cost is much greater, collecting the basic data may ease the subsequent decision on appraisal.

Most fields require one or more appraisal wells to build confidence before investment. In the early days of North Sea exploitation, before modern 3D seismic was available, after making a discovery, several appraisal wells would be drilled to delineate a field. In one case, a well was drilled to the north and south of the discovery and a platform located in the center: the first well drilled to the east crossed a fault that put the reservoir into the aquifer. This near disaster was corrected by locating a second platform 2 km to the west so the whole field could be successful accessed; this was before the days of extended reach wells.

An appraisal program should meet three criteria:

1) Information gathered must be relevant and have the potential to change your belief about a given uncertainty.
2) The data must have a material impact on decision-making.
3) The cost of acquisition must be less than its value (VOI – value of information).

Additional 2D/3D seismic acquisition can be expensive if required for onshore appraisal, and drilling may be less costly. Drilling a well may allow the acquisition of additional core data, well logs, or drill stem tests.

The commonest way of looking at these criteria are to develop decision trees to address the different possible outcomes and their impact on the development. Ideally, appraisal will move smoothly into development planning, and thus reduce the cycle time from discovery to first production. A development team will be put together combining subsurface professionals, facilities engineers, commercial specialists, economists, and legal experts to review each aspect of the proposed project.

Glenlivet Field, West of Shetland *A successful exploration well tested a seismic anomaly in 600 m of water during the summer drilling season (Horseman et al. 2014). Updip of the well the seismic response became tuned, reducing confidence in the in-place gas volume. The vertical well was fully logged and cored over a third of the reservoir sands. Given good weather and time on the rig schedule the operator decided to drill an updip sidetrack to reduce the seismic uncertainty; the sidetrack confirmed a thick reservoir section to*

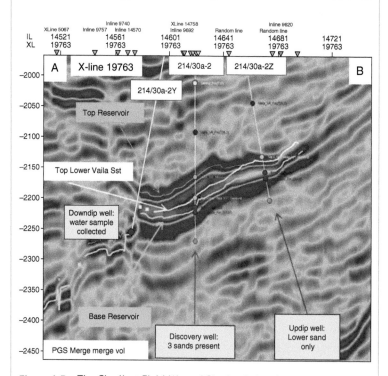

Figure 1.3   The Glenlivet Field, West of Shetland, showing the seismic response of the reservoir plus the discovery well and two sidetracks that were drilled to fully appraise the field prior to development. *Source:* Courtesy of DONG Energy.

the edge of the anomaly, confirming the optimistic gas volume estimate. To gain access to an export pipeline a sample of the downdip aquifer was required; this sample would represent any produced water. So the operator decided to drill a second sidetrack and recover a water sample. Each sidetrack was logged to establish a reservoir correlation. There was no requirement to test the well because the core and log data confirmed a high deliverability reservoir; the gas in place could be drained by one well alone, and no further appraisal was required (Figure 1.3).

From these results a development plan was written that required the drilling of two vertical wells, each with sand screens to minimize sand invasion and downhole gauges to monitor gas production. Two new wells were planned so that neither well was over produced and there would still be a producer should either well fail. The ultimate production rates from the field would be dependent on the available capacity in the pipeline.

1.3 Volumetrics

The starting point for any field development is an estimate of the in-place hydrocarbon volume. Is there enough oil or gas to make a development economically viable or do the hydrocarbons remain future resources and not reserves?

Conventional volumetric calculations to estimate hydrocarbons initially in-place (HIIP) require a number of simple input parameters, most of which are in the remit of the petrophysicist. The input parameters are related in the following way:

Hydrocarbons initially in-place:

$$HIIP = \frac{GRV.NTG.\phi.\left(1 - S_w\right)}{B}$$

where:

GRV	=	gross rock volume
NTG	=	net : gross
ϕ	=	porosity of net reservoir
$(1 - S_w)$	=	hydrocarbon saturation
B	=	formation volume factor

Gross rock volume (GRV) is the volume of rock between the top reservoir structure and the closing contour or hydrocarbon water contact.

Net : gross (NTG) is that part of the reservoir containing moveable fluids. In a 3D model this can be calculated from a facies model or by the application of a series of cut-off values based on an effective porosity and/or permeability. The ratio of NTG to GRV gives the *net rock volume* (NRV).

Porosity is the capacity of a rock to store fluids; the pore volume of a reservoir model is the total of all cells with an effective porosity and is directly linked to the NRV. Cells with effective porosity may be defined by a facies model or by some cut-off value.

Hydrocarbon saturation is the proportion of the pores filled with hydrocarbon rather than water; the volume of hydrocarbon plus the volume of water will be unity, $S_h = (1 - S_w)$, where S_w is the proportion of water in the pore space. Only those cells that have been designated as being NRV will be counted in the summation to give the reservoir hydrocarbon volume.

Formation volume factor accounts for the increase in hydrocarbon volume between the reservoir and the surface; this is a function of the change in pressure between reservoir and surface conditions and depends on the fluid description.

There are a number of dynamic methods used to estimate HIIP such as decline curve analysis, material balance, or reservoir simulation; these will be discussed further in Chapter 5.

1.4 Drive Mechanism

Oil and gas reservoirs have a natural drive mechanism that provides the energy to move the hydrocarbons toward a producing well. The common natural drive mechanisms are water drive, gas expansion, solution gas drive, compaction drive, and gravity drainage. One drive mechanism initially dominates production, but a combination of mechanisms often occurs over the period of field life. Depending on the drive mechanism it is possible to estimate potential recovery factors ranging between 5 and 30% for oil fields and up to 80% for a gas field that depletes naturally (Table 1.1). To improve recovery for any field normally requires a secondary method to maintain the energy in the reservoir.

Table 1.1 Recovery factors for different fields under differing production methods. Rules of thumb!

Recovery factor	Low (%)	Medium (%)	High (%)
Oil field – by natural depletion and without aquifer support	10	15	20
Oil field – by efficient water or gas injection or strong aquifer support	30	50	60
Gas field – by natural depletion	60	70	80

1.5 Field Development

Having delineated the field and acquired the necessary reservoir and production information a development plan can be written. There will always be uncertainties in the reservoir description that will need to be considered in the development plan. The key elements are the number of wells required to drain the reservoir and the surface facilities needed to fulfill any contractual or legislative agreements. In the situation where gas is contracted on an annual basis it is possible to plan a simple rate of production per well to fulfill an agreed annual decline profile. However, if the gas market is liberated, such as in the UK, each well or group of wells is required to deliver production at hourly or daily rates to fulfill the needs of commercial and domestic users. In this latter case, having wells that can be switched on and off like a tap become essential. Oil production is managed on a different monetary basis where value is made on delivery to the refinery and the quality of the oil.

Drilling onshore is usually done on a regularly spaced pattern: initially widely spaced to allow for subsequent infill drilling or water-flooding; these might be in the original plans or form a secondary phase of recovery. Offshore field developments require higher CAPEX including large platform structures and pipelines or storage facilities. The number of wells is often restricted by the surface facilities and wells must be engineered to maximize access to the reservoir.

Field developments are often staged or phased: a subtle distinction where *staged* plans are approved at time of commissioning whereas *phased* plans are based on the results of the initial development. In the past, many offshore fields were over engineered or displayed 'gold plated' solutions to basic issues. Those days are long gone, and every nut, bolt, and piece of pipe must be justified during the economic assessment of a project.

Hyde Field, Southern North Sea: *Planning a new field development is a matter of looking at the potential resource and deciding how best to monetize that resource. When the UK gas market was liberated in the 1990s there was a 'dash for gas' by those companies with export infrastructure. Small fields that previously were uneconomic under fixed*

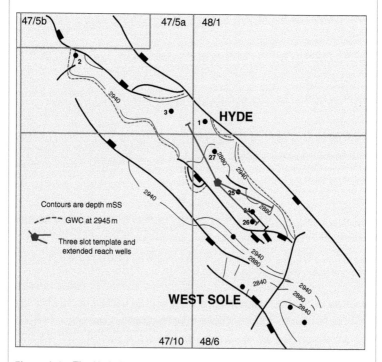

Figure 1.4   The Hyde Field, UK Southern North Sea, showing the main structural features plus the locations of the three extended reach wells drilled from a three-slot subsea template.

contracts suddenly became swing feeders to the existing system: wells that could be switched on and off as required. The Hyde Field was discovered in 1982 but was only brought to production in August 1993 when the economic risk could be mitigated through new technology: extended reach near horizontal wells (Steele et al. 1993). Extensive appraisal of a similar nearby accumulation indicated a number of important constraints on development: a partial gas column (~105 m); low permeability (1 mD mode); high water saturation (40%); low, natural flow rate (12 MMscf/d): effectively the field was in a transition zone. This information indicated that production wells would require some form of stimulation to achieve economic rates. Because of the partial gas column, horizontal drilling was chosen over hydraulically fractured deviated wells (Figure 1.4).

To demonstrate that the field could be best exploited by this method, a pilot well was drilled and cored at the crest of the structure prior to drilling a horizontal sidetrack. The sidetrack was drilled and steered using LWD technology to maintain maximum penetration of the best quality reservoir layer: ultimately 420 m of productive reservoir was penetrated. The well was tested using a predrilled slotted liner to allow future use as a producer. The well tested at a rate of 69 MMscf/d, which was a significant increase on previous vertical/deviated wells. A second horizontal production well was drilled with similar results and the field was developed with a three-slot subsea template and a not normally manned platform.

1.6   Reservoir Simulation

These days, no development manager would consider presenting a project to an investment board without the results of a reservoir simulation to support the assumptions made. Reservoir simulation is a numerical method of representing flow in a reservoir toward the production wells and on to the surface under a number of chosen scenarios. The model can be used to run production forecasts and for assessing uncertainty for each development scenario.

The numerical model usually comprises an upscaled geological model, fluid data (pressure–volume–temperature [PVT]), and relative permeability information and fluid distribution: these form the initial state model prior to running any simulation. Well completion and production information is required plus some knowledge of the surface facilities at least at the wellhead. From these initial simulation runs it is possible to predict rates of production, pressure decline, and ultimate recovery for a given scenario, thus determining the value of the development. Ultimately, the model can be used to 'history match' production and then for prediction. Like all models the results are nonunique and dependent on the quality of the reservoir description. The Appendix has a simplified description of the process of reservoir simulation.

1.7 Field Production

The facilities required to handle the produced hydrocarbons depend on the type of hydrocarbons and the location of the reserves: offshore or onshore. An onshore oil field can be produced to local storage tanks prior to truck or pipeline transportation to a refinery. Onshore gas may be piped directly into local energy generation plant or liquefied for transport. Offshore platform developments for oil and gas usually require a pipeline to an offshore or onshore gathering system for further processing. Further offshore, and in deeper water, floating production and storage vessels (FPSO) are often a preferred solution.

A large part of field development planning addresses these fundamental options and the economic benefits and consequences of one or the other. Before committing to a major deepwater gas development the partner companies will construct a risk assessment based around three options: a deepwater installation to process the gas prior to export, a shallow water processing platform and pipeline, or a subsea development and a pipeline to an onshore gas processing plant. The subsea development and pipeline to shore was selected in part because it allowed for significantly greater processing capacity and opportunity for future gas discoveries to be developed through this infrastructure.

Magnus Field, Northern North Sea: *Magnus was discovered in 1974, but first oil was not until 1983. The field has an estimated 1.54 billion barrels of oil in-place with reserves of around 870 million barrels: around 56% recovery. The field was developed through a single drilling and production platform: the largest single piece steel structure in the North Sea. Initially there were facilities for 17 oil production wells, five water injection wells, and nine spare slots, and the production capacity was 140 000 BOPD and 2.5 MMscf/d of gas. The oil is exported via a 91 km pipeline to the Ninian Central Platform and then to the Sullom Voe terminal on Shetland. Produced gas is exported to Brent A via a 79 km pipeline and onwards through the FLAGS pipeline system to St. Fergus gas terminal on the Scottish mainland (Shepard 1991).*

A major project to increase recovery and extend field life was proposed in 2000 and implemented three years later. The project required importing gas from two West of Shetland oilfields via a pipeline to Sullom Voe where liquefied petroleum gas was added before being delivered by another pipeline to the field where it was reinjected into the Magnus reservoir to aid pressure support and hence recovery as a water–alternating gas (WAG) project. It was expected that an additional 50 million barrels would be recovered, and the field life extended by a further 5–10 years. The whole IOR project cost around £320 million. Ten years on, another field rejuvenation project was initiated to maximize the ongoing depletion strategy and to build a better understanding of the plant and well reliability both of which were successful in managing to achieve planned voidage replacement and plant restoration. The field operating efficiency has grown from 40% in 2013 to over 70% in 2017. At the end of last year, the company estimated remaining 2C resources at around 50 MMboe, but it has also identified various drillable targets that could lift remaining mobile oil in-place to 270 MMbbl. These, and a potentially large gas discovery made by the former operator, could form the basis of a long-term, low-cost development program.

1.8 Reservoir Monitoring and Surveillance

Monitoring the key reservoir indicators of hydrocarbon production is essential for both reserve estimation and management during the field life. Essentially, oil or gas production on a well-by-well basis, or by gathering station and by total field production, is required for basic economic valuation: the value of how much oil or gas is produced is the reward for the investment. How much water or gas is injected is monitored to estimate the pressure maintenance strategy, and the pressure decline in each well or reservoir unit may also be recorded to determine the wells that might need more support. Modern methods of reservoir monitoring include permanent gauges to measure real-time oil, water, and gas flow, sending the data to central data-gathering centers for analysis and for further intervention. This area of reservoir management is where many of the greatest advances have been made in recent years.

1.9 Improved Hydrocarbon Recovery

As hydrocarbons are produced the reservoir loses energy due to a decline in pressure, unless there is an active aquifer helping to support production; when this not the case it is necessary to provide that extra boost to the reservoir. This is commonly done through water injection into the aquifer, or gas injection into the crest of the field; gas compression may be a means to extending the life of a naturally depleting gas field. All of these methods require additional expenditure at some point in field life; sometimes this is built into the original development plan if appropriate. The economics must drive these decisions; there must be sufficient added value in extending field life, but the inventiveness and imagination of petroleum engineers never ceases to amaze me.

Beryl Field, Northern North Sea: Brought on stream in 1976, and designed for a 25 year field life, the Beryl Field is still a location for near-field exploration and production over 40 years later. At a peak production rate of 200 000 BOPD an early phase of satellite developments slowed down the subsequent decline, however by 2012 the end was in sight. Following the acquisition of 3D broadband seismic in 2012–2013, PSDM seismic processing of the data has resulted in a better understanding of the structural and reservoir complexity. In view of this, a portfolio of near-field exploration targets was drilled in 2015–2016 with four commercial discoveries in two years, a 66% success rate. A key element of these wells was the planned sidetrack to evaluate different structural elements: one sidetrack penetrated a 200 ft column of oil that was greater than the original Beryl oil column. Production from these wells started a year later and, because of the existing infrastructure and its capacity, improved their economics significantly and halted the production decline. These results have arrested the production decline and extended field life by a further 5–10 years (Helgeson et al. 2019).

1.10 Cessation of Production: Field Abandonment

All oil and gas fields will eventually cease to be economic and must then be abandoned incurring a final cost to the overall CAPEX: hopefully this will be after long-term financial growth of profits. Around the world there are many oilfields still active after 50–100 years of production; indeed, many have long-term investment plans to continue their profitable life. However, some fields have had very short lives because of unforeseen circumstances: usually a matter of under-appraisal or having been under managed.

An example of such would be a trio of gas fields in the Southern North Sea that were abandoned after less than 10 years of production because of an unexpected influx of highly saline formation water that effectively blocked the pore space around the wellbore reducing gas flow completely. The fall in pressure toward the wellbore resulted in salt coming out of solution and crystallizing. Some remedial steps were taken to flush out the salt, but these were ultimately unsuccessful and the fields ceased production well before the expected ultimate recovery was achieved (Ketter 1991).

1.11 Summary

Developing and managing an oil or gas field requires a multidisciplined approach that can address the multiplicity of technical, commercial, and environmental issues. Hopefully, the rest of this book will demonstrate some of the challenges faced by oil companies, their staff, and stakeholders.

2

Reservoir Management Process

The oilfield life cycle typically follows a well-trodden path from discovery, through appraisal to development, production, and finally abandonment, and at every stage data is collected that guides the investment decision-making (Figure 2.1). In the Western world the economic driver is simply the ultimate recovery: is it worth investing money for a quantifiable return? Are there a given number of wells to be drilled after which the investment shows a declining return? After making a discovery of oil or gas the next question to be asked is how much hydrocarbon is present and what is required to make an economic development feasible. Many offshore discoveries lie undeveloped until the economic situation changes: a rise in the price of oil or some new infrastructure built to recover a stranded asset. The process of developing an offshore asset may be very different to the onshore situation because of the greater initial investment required and the risk associated with a less well-defined project having more uncertainty. This section focuses primarily on the challenges of developing an offshore field.

Before looking at field appraisal or delineation it is worthwhile asking the question, 'why did we drill the discovery well here?' Today the answer will usually be 'because of the seismic bright spot' or 'because of a change in amplitude': the interpretation of seismic data drives exploration decisions. In the past, exploration wells would be drilled on a structural high or possibly because of a faulted section in the stratigraphy. Whatever the

Reservoir Management: A Practical Guide, First Edition. Steve Cannon.
© 2021 John Wiley & Sons Ltd. Published 2021 by John Wiley & Sons Ltd.

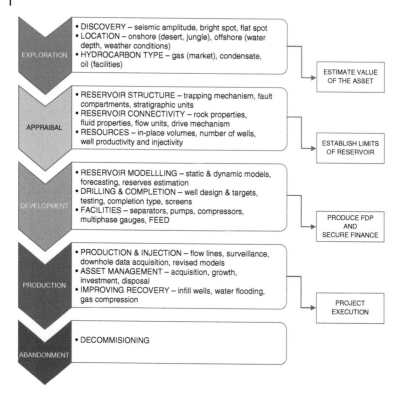

EXPLORATION
- DISCOVERY – seismic amplitude, bright spot, flat spot
- LOCATION – onshore (desert, jungle), offshore (water depth, weather conditions)
- HYDROCARBON TYPE – gas (market), condensate, oil (facilities)

ESTIMATE VALUE OF THE ASSET

APPRAISAL
- RESERVOIR STRUCTURE – trapping mechanism, fault compartments, stratigraphic units
- RESERVOIR CONNECTIVITY – rock properties, fluid properties, flow units, drive mechanism
- RESOURCES – in-place volumes, number of wells, well productivity and injectivity

ESTABLISH LIMITS OF RESERVOIR

DEVELOPMENT
- RESERVOIR MODELLLING – static & dynamic models, forecasting, reserves estimation
- DRILLING & COMPLETION – well design & targets, testing, completion type, screens
- FACILITIES – separators, pumps, compressors, multiphase gauges, FEED

PRODUCE FDP AND SECURE FINANCE

PRODUCTION
- PRODUCTION & INJECTION – flow lines, surveillance, downhole data acquisition, revised models
- ASSET MANAGEMENT – acquisition, growth, investment, disposal
- IMPROVING RECOVERY – infill wells, water flooding, gas compression

PROJECT EXECUTION

ABANDONMENT
- DECOMMISIONING

Figure 2.1 Stages in the life of an oil or gas field from exploration to abandonment highlighting the key data analyses and outcomes.

reason for drilling an exploration well, it should be possible to determine the likely closure of the field and thickness of the reservoir zones, and if the well has penetrated the water leg, then a base level is established. From this, using a relatively simple area–depth map and planimeter, an estimate of the hydrocarbons in place can be made, if an average porosity and hydrocarbon saturation is 'known'. This initial estimate of resources is often a robust benchmark used for future calculations of hydrocarbons in place using more sophisticated methods. As more information becomes available on the ranges associated with these properties minimum, most likely, and maximum values can be reported. An estimate of the ultimate recovery is dependent on many other factors.

2.1   Field Appraisal

Field appraisal is the post-discovery evaluation of a potential development opportunity. Put simply, what data is required to appraise the size
and economic value of an asset? How many wells might be required to
develop the asset and what is the commercial driver? Appraisal costs
generally fall in to two categories: additional seismic acquisition and/or
processing, and additional drilling. There may also be a requirement for
coring or well testing, both of which incur costs during acquisition and
subsequent analysis. Whatever steps are taken the key objective is to
identify the optimum development scenario and to ensure that project
economics are robust; this is done by minimizing pre-development costs.
You will often hear the term 'value of information' or VOI: what positive
outcome can be achieved by spending money. This is often the approach
of engineers and accountants whereas geoscientists always see value in
more information!

To fully characterize an oil or gas field the following subsurface data
should be acquired:

- Processed and interpreted 3D seismic data to provide a mapped structural surface of the top reservoir and quantified reservoir property and
  fluid information.
- Core data for sedimentological and correlation studies, and rock property information: porosity, permeability, and grain density are required
  to constrain wireline log analysis.
- Wireline logs to give continuous petrophysical analysis of porosity,
  fluid distribution, and producibility of an interval, together with indications on reservoir correlation.
- Wireline pressure measurements and fluid samples through the main
  reservoir zones.
- Transient well test analysis and long-term production tests to gain well
  and reservoir productivity estimates.

2.1.1   3D Seismic Data

The acquisition of additional 3D seismic data or its reprocessing is
required primarily for the identification of reservoir units and faulted
compartments. In the past, structural and stratigraphic complexities

have been missed during the development planning phase resulting in numerous reservoir management problems ranging from the need to drill additional production wells or poor connectivity between producer wells and injector wells. Today, however, the use of quantitative seismic techniques such as seismic inversion, rock physics analysis, and various signal-stacking methods in time and depth have enabled geophysicists to increase their impact on reservoir characterization. The key input of the geophysicist to successful appraisal and development is still in accurate depth conversion based on sufficient well-tie data to accurately define the top reservoir structure over the accumulation: around 30% of the resource uncertainty is associated with depth conversion issues. Seismic data provides low resolution but high-density data coverage over the area of interest unlike well data, where the resolution is generally greater, but the coverage of data is low because of the cost of drilling.

2.1.2   Core Data

You will often hear drillers say that the only thing you can be sure of about core data is that at least you know where it comes from: not a view that geoscientists would agree with given problems with recovery! At the basic level whole core or sidewall cores can provide information on the lithology and rock properties of the interval cored. Core data provides the main input for calibration of seismic and wireline log data, notwithstanding the physical changes that occur to the rocks and fluids during the recovery process. Core data also allows sedimentologists to argue about the depositional environment represented by the reservoir: terrestrial or marine; aeolian or fluvial; shallow or deep marine? While cuttings data collected during drilling will indicate the major lithological and stratigraphic boundaries, whole core data can demonstrate the nature of these boundaries; they may be gradational or sharp based, erosional or faulted.

2.1.3   Borehole Log Data

Wireline logs and real time logging while drilling (LWD) data give a continuous record of the response of the target reservoir to electrical, sonic, and nuclear stimuli. Some measurements are in response to rock properties such as lithology or porosity, others to the fluids present in the reservoir pore space. The continuous nature of the data shows the large-scale variation in the overall depositional sequence, as well as the smaller-scale interfingering of beds, leading to a stratigraphic hierarchy of reservoir units. All of the measurements made by wireline or LWD tools are indirect and require significant processing and interpretation before they can be used to characterize the reservoir rocks or the fluids. Only with the addition of direct sampling through core or formation testing can these data be confirmed: 'petrophysics without core data is just log analysis'.

One special type of wireline tool provides specific reservoir information on fluids and pressure and that is the 'wireline formation tester' (WFT, RFT, MDT). This tool is designed to measure the reservoir pressure and to collect a fluid sample at one or more specific depths in the reservoir. A series of pressure measurements over a large interval can be used to generate a pressure gradient that may be used to identify vertical compartments or separate reservoir units.

2.1.4 Well Test Data

Transient well tests are carried out in appraisal wells to confirm what hydrocarbons will flow, the rate of production, and rate of pressure decline due to production. Other information, such as drainage area and the presence of boundaries to flow, may also be identified from the analysis of the results; however, the interpretations are nonunique and open to much misinterpretation.

Well tests are designed to specific patterns to include flow periods and shut-in periods of 6–12 hours usually, but longer term, extended well tests (EWTs) are sometimes given approval by partners and governments if a speedy development cycle is planned. An added advantage of an EWT is that the oil produced can be stored and sent to market: three months production might pay for the cost of the well!

A fully appraised field should be ready to move to the next stage, which is development planning. But how often is a field developed that doesn't meet expectations because of under appraisal or, more likely, misinterpretation of the data? The Buzzard Field is one of the more recent exploration successes in the Central North Sea. The field was discovered in 2001, encountering a 400 ft gross oil column in Upper Jurassic turbidite sands. A sidetrack of the discovery well extended the gross oil column to

750 ft and the field laterally by 4400 ft. The hydrocarbons are stratigraphically trapped at between 7800 and 9175 ft below mean sea level. After the development plans were approved and during development drilling, it became apparent that there were high concentrations of hydrogen sulfide (H_2S), up to 14%, in the hydrocarbon stream; an unexpected event that might be explained by bad luck or bad analysis. Either way a fourth platform was required to strip out the H_2S.

Schiehallion Field, West of Shetland: *The field was discovered and delineated using a 3D seismic survey acquired and processed in 1993 as part of a 2000 km² regional survey designed to illuminate potential targets. The first well drilled was 204/20-1 that targeted a seismic amplitude anomaly based on a strong amplitude versus offset (AVO) response associated with hydrocarbon bearing Paleocene sands. A subsequent sidetrack, well −1Z, was drilled to acquire further data to better define the accumulation. The discovery well penetrated an oil–water contact (OWC) at 2064 m true vertical depth subsea (TVDSS). Four further appraisal wells were drilled in 1994 to better delineate the field. A high angle/horizontal well, 204/2-5, was drilled in 1995 that penetrated ~1700 m of gross reservoir and was completed for an EWT that proved the deliverability of the reservoir producing at a rate of ~18 000 BOPD; the well was suspended as a future producer. A near crestal vertical well was also drilled in 1995 to eliminate the possibility of a gas cap. This appraisal program determined recoverable reserves of 340 MMbbls. In 1996, after project sanction, a high-resolution baseline seismic survey was acquired with repeat surveys acquired every two years since the start of production; this has been a major tool in reservoir management. (Leach et al. 1999)*

2.2 Field Development

Field development planning is based on passing a series of 'decisions gates' or 'milestones' (Figure 2.2). At each gate a stop–go decision is made by the whole team, either to step back to gather more data or proceed to the next stage of the plan. Field development begins where the exploration phase ends: in other words, once a decision is made to

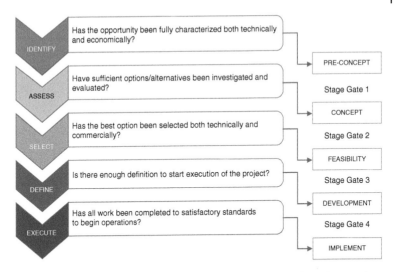

Figure 2.2 Stage gates in a development plan and the questions that need to be answered before progressing through each stage.

appraise the discovery. The first decision is taken when the opportunity has been identified and framed technically and economically; this is made after appraisal is completed. Before moving forward to the next stage, the question is asked whether or not all the possible options have been investigated and evaluated. Typically, a review of the estimated resource in place is made and a value placed on that estimate allied with the economic return. Ask the question: 'Where and when will what volumes of which fluids will be produced?' This is where risk and uncertainty come in to play: what can be done to reduce the uncertainty and limit the acceptable risk?

Key factors in development concept selection are:

- Location: Proximity and location of existing infrastructure (export options), water depth, climate (hurricanes), areal extent of the field, legislation, environment, and politics.
- Volumes: Key issue to design facilities to allow for all uncertainties in development.
- Fluids to be produced: Must allow for variations in time in the process design, materials selection, production handling, injection or compression capacity, and export options.

- *Productivity*: Understand how and when the field will be produced. Well numbers, flow rates, GOR over time, pressure, and maintenance requirements.

An analysis of these factors usually leads to a number of development concepts from which the final plan will be selected. The selection may be a simple solution to a single project or made with a more strategic view of building infrastructure for multiple projects. Ask the question, 'has the best concept been selected both technically and commercially?' A development concept may be uncertain because we are unable to identify a realistic development option with a reasonable degree of certainty.

When there are multiple options the range of reserves is wide, but each option is essentially suited for a particular field size and hydrocarbon offtake. The aim of the appraisal is to eliminate the various development options by acquiring additional information. The idea is to narrow the reserves range and settle on a single development selection. Reducing the time between discovery and first oil can significantly increase project value (net present value – NPV), provided risks and uncertainties are properly identified and managed. For fast track development, it is critical that all stakeholders are fully aligned (partners, government, and suppliers). Fast track projects need early and clear signals from management to prevent roadblocks. It is worth remembering that commercial and political pressure can often compromise field development.

A development has reached economic maturity when one of the following decisions are achieved:

- *Uneconomic*: Project fails to make an acceptable return from any scenario. No further appraisal justified.
- *Feasible*: A discovery may be economically feasible if it makes an acceptable return from an upside scenario, even though the most likely case is uneconomic. Further appraisal may be sanctioned.
- *Viable*: A discovery is viable if the most likely case makes an acceptable return, even though downside remains uneconomic.
- *Robust*: A discovery makes an acceptable return under any scenario.

A discovery may be economically robust under one development option even though other options are uneconomic. Should this be the case, further appraisal is justified. Further appraisal of a robust project may be justified if improvement in value of the project is greater than the cost of continued appraisal. Economics of a project may change due to changes in technology or changes in commodity prices.

Data acquisition during development may depend on what has already been collected during exploration and appraisal. If sufficient core material has been collected to fully characterize the geology of the reservoir and to calibrate the petrophysical properties, either measurement-while-drilling (MWD) or wireline logs may be enough to progress the project.

Schiehallion Field – West of Shetland: *During development significant geological and petrophysical information was collected that improved the description and understanding of the reservoir. Three main Tertiary submarine fan sandstone bodies were recognized separated by non-reservoir mudstones. The sandstone packages comprise a complex of channel and interchannel facies. The sandstone porosity varies between 25 and 30% with associated permeability of 800–1600 mD. The net : gross ratio (NTG) of the channel bodies is around 70% while the interchannel NTG is 30–50%.*

Key elements of the Schiehallion development plan were:

- *The use of horizontal wells.*
- *The use of advanced seismic technology for well positioning to maximize production.*
- *A phased and flexible drilling strategy.*
- *Provision of sufficient water injection to completely replace oil production.*
- *Use of gas lift in all production wells.*
- *Disposal of surplus gas at a remote injection site.*

The sandstone bodies are well resolved on 3D seismic allowing development to be geosteered using seismic attributes as a guide. Even though the initial appraisal of the well suggested that reservoir sands were well connected, during development drilling and early production it became apparent that sand bodies were more isolated, requiring targeted drilling of producer–injector pairs. The poor well performance was recognized as rapidly declining bottomhole flowing pressures (BHFPs), declining production rates, and rapidly rising gas–oil ratios in 50% of the production wells. Injector wells also showed rising tubing head pressures (THPs) and reduced injectivity (Leach et al. 1999).

2.2.1 Field Development Plan

What should be included in a field development plan (FDP)? The UK Oil and Gas Authority has published an outline contents list for an FDP comprising three main sections: an executive summary, the field description, and the development and management plan, which is very similar to the outline of this book (Table 2.1).

There are two major costs associated with field development; first, construction of the main production facilities and, second, drilling. Onshore development costs are generally less than offshore, and this applies to drilling as well as facilities. Primarily the number of wells will determine onshore drilling costs; offshore, the number of wells is generally fewer but may be significantly more expensive.

Table 2.1 Main elements of a field development plan (FDP) document.

Section 1. Executive summary

Section 2. Field description

2.1 Seismic interpretation and structural configuration

2.2 Geological interpretation and reservoir description

2.3 Geological model

2.4 Petrophysics and reservoir fluids

2.5 Hydrocarbons initially in place

2.6 Reservoir modeling approach

2.7 Reservoir development, improved and enhanced recovery processes

2.8 Wells design and production technology

Section 3. Development and management plan

3.1 Preferred development plan, reserves and production profiles

3.2 Drilling and production facilities

3.3 Process facilities

3.4 Project planning

3.5 Decommissioning

3.6 Costs

3.7 Field management plan

Section 4. References

Source: Oil & Gas Authority (2018).

Onshore wells are often drilled to a predetermined pattern per acre or a given well spacing; most wells will be vertical or S-shaped if a number are drilled from the same well pad. Typical patterns will be five-spot or nine-spot with a spacing of 250 m or 500 m. The areal extent of a field will often determine the number of wells required to effectively drain the accumulation. Starting with a wide well spacing allows for infill wells later in field life.

Offshore, wells are usually drilled from a platform and a predetermined number of slots are provided in the wellhead template. Subsea developments will usually have a smaller number of wells from a template, three to five slot being common, and additional single wellhead sites for step-out wells with these single wellheads being tied back to the main site. The cost of the wells can easily be 10 times more than for a completed onshore well; this is because of higher rig rates, deeper, longer wells that require complex well designs and completions.

Once into the field development stage the main decision gates are about concept selection; whether to go with a fixed platform, floating vessel, or a subsea development. Additionally, whether the hydrocarbons will be exported via a pipeline or tanker, and what level of processing will take place on site. This work is usually called front end engineering design (FEED) after which a final investment decision (FID) will be made such that the project moves to the next stage. Most fields will be developed with a number of wells drilled pre-production so that a cash flow stream can begin as soon as possible.

After the decision is made to proceed with a development then a whole new project life cycle comes in to play, that of the facility engineers (Figure 2.3). The key to project management can be summed up in three small words: cost, schedule, and quality. A successful project can be measured by being on budget, on time, and fit for purpose. To judge this, a series of performance tests are introduced upon completion that demonstrate that the project has achieved the expected results when the project is handed over to operations.

2.3   Field Production

Bringing a field on stream represents a significant achievement for the operator, partners, and other stakeholders, but especially for the teams of professionals involved. Many fields are developed in stages, sometimes pre-planned and sometimes as a result of new information or changes in

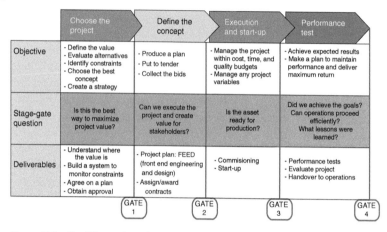

	Choose the project	Define the concept	Execution and start-up	Performance test
Objective	- Define the value - Evaluate alternatives - Identify constraints - Choose the best concept - Create a strategy	- Produce a plan - Put to tender - Collect the bids	- Manage the project within cost, time, and quality budgets - Manage any project variables	- Achieve expected results - Make a plan to maintain performance and deliver maximum return
Stage-gate question	Is this the best way to maximize project value?	Can we execute the project and create value for stakeholders?	Is the asset ready for production?	Did we achieve the goals? Can operations proceed efficiently? What lessons were learned?
Deliverables	- Understand where the value is - Build a system to monitor constraints - Agree on a plan - Obtain approval	- Project plan: FEED (front end engineering and design) - Assign/award contracts	- Commisioning - Start-up	- Performance tests - Evaluate project - Handover to operations
	GATE 1	GATE 2	GATE 3	GATE 4

Figure 2.3 Facility engineering approach to project planning to ensure quality, cost, and schedule.

the economic climate. A phased development is one where the decision to move to a second or third phase is dependent on the results of a preceding phase. To this end it is essential that production data and pressure information are captured on a daily basis at a minimum. This is where the greatest advances in technology have occurred in recent years as smart wells and smart fields have become the accepted norm in the industry.

Permanent reservoir monitoring technology enables operators to increase safety and oil recovery through monitoring the reservoir's behavior and environment. Reservoir and production monitoring and control devices are critical elements for better reservoir management. The increasing use of topside and subsea metering systems provides important information on the production characteristics of the reservoir. All these devices can now be logged and controlled remotely and in real time. As a consequence, it is now technically possible to manage fully the reservoir exploitation process automatically from a remote location.

Another area of monitoring a field during production is the use of time-lapse seismic: routine acquisition of 3D seismic every two to four years. In this approach a baseline survey is taken before production starts and repeated over the production lifetime of the field. Differences in the seismic amplitude or other attributes between the surveys can reveal the movement of fluid contacts or areas that have been drained. Increasingly, seismic monitoring can be achieved by permanent ocean bottom cable (OBC) and ocean bottom node (OBN) arrays reducing the cost and improving the results. Timing of the repeat surveys is important to maximize the benefits for reservoir management so they can have an impact on the location of an infill well or phasing of the development. Processing of the seismic should also be considered in the timing of any repeat survey, because it will take significant time before it may have value.

There are many aspects of field production methods to consider: from drilling and completion strategies to pressure support and artificial lift mechanisms. Drilling offshore from a fixed platform means there will be a limit to the number of wells that may be drilled. The giant Brent Field in the Northern North Sea was developed with four platforms with a total of 154 well slots; of these 25% were used for water injector wells. These wells were all deviated up to ~35° from the vertical increasing the total well length significantly. By the end of the production phase, a total of 220 platform wells and three subsea wells had been drilled; many slots being reused as wells ceased production. Of comparable size and economic importance, the Forties Field was developed with five platforms and a total of 142 well slots of which 103 slots were utilized. With a large and active aquifer the need for water injection was minimized.

Production wells may have a range of completion types from a simple 'barefoot' to a slotted liner or more sophisticated gravel pack and mesh equipment where there is a high risk of sand production. Where it is difficult to access a production well, or there should be a requirement for workover activity, a 'belt and braces' approach to completion is often proposed. Deciding what type of completion should be installed often requires rock strength tests to be made on core taken from the reservoir. These decisions should be made during the development discussions.

Monitoring production and injection rates, together with regular pressure measurements, are the minimal requirements of data acquisition for reservoir management. Today real time monitoring and smart well technology is being implemented in many new developments around the world. The additional cost of hi-tech solutions must be compared against the long-term value of their installation, and offshore this is easier to justify than onshore.

2.4   An Integrated Team Structure for Reservoir Management

The advantages offered by working in an integrated team for communication, project management efficiencies, and ultimate results, both technical and commercial, cannot be stressed enough: a well led team with definable objectives should always be a more effective organization than individuals working alone or a traditionally structured organization. A good team has the following drivers:

- A shared vision of the ultimate objective – maximizing productivity.
- High levels of trust between individual members.
- A clear definition of quality and a means of measuring it.
- Realistic goal setting and measurement.
- Group rewards tied to performance.

A team will perform well when there is synergy between members, and tasks that generate meaningful challenges will lead to better problem solving. Teams generally fit into two functional categories: *permanent teams* and *task forces*, which are structured either as *hierarchical organizations* or as a *peer group* with leaders who are often appointed or evolve (self-selecting) for different phases of a project.

There are a number of ways that different companies have structured their teams depending on company culture, country of operation, or project objective. The benefits of any approach should outweigh the costs of doing work in a different way, including costs associated with the change process. The following list shows examples of the ways in which companies have chosen to operate:

- *Employee Teams*: Traditional in-house multidisciplinary teams often organized in an hierarchical structure.
- *Partnerships*: Formal arrangements with another company to jointly undertake a project.
- *Strategic Alliance*: Usually with a service provider who delivers a solution based on the company vision.
- *Outsourcing*: Commissioning a part or the whole solution through a third party.

With respect to reservoir management, both strategic alliances and outsourcing of projects have become increasingly a feature of company workflow. Both models lead to closer working relationships between companies and service providers with the expectation of reduced costs and more efficient work practices designed to limit duplication of effort. For these operational changes to be successful a strong partnership or alliance must have:

- A joint vision or alignment of purpose.
- A period of detailed planning before implementation.
- Management commitment and team acceptance of the partnership.
- Strong complimentary skills between partners and team members.
- Clearly defined roles and responsibilities.
- Commitment to the long term – at least the expected length of the project.
- Common corporate and personal aims and objectives – a common culture.
- Well-developed interpersonal skills among the team and its leaders.

The changes that are needed to successfully move to a team-based work culture require that the management structure also change. The traditional hierarchical structure is dependent on close control of strategies and tasks from above: both functional and operational micro-management. The new approaches demand that the team is charged with managing their tasks such that they are in line with the greater strategy set by the upper management and may even have had an influence in their design.

A successful multidisciplinary team manages production for a field and looks for ways of getting more hydrocarbons out of that field. In a large company, this team will include some combination of the job titles listed in the Table 2.2. Teamwork is essential because the staggeringly complex nature of a subsurface operation means that the various disciplines have to integrate their specific areas of expertise for the venture to be successful (Figure 2.4). Some oil companies have separate geology and engineering departments, but regardless of the make-up of the team, short lines of communication should exist within the team such that an inclusive atmosphere of shared purpose is created. Any problems that arise can then be quickly recognized and solved by common directed action.

In an earlier book on reservoir modelling (Cannon, 2018) I discussed some of the topics presented at a conference titled "Recognising the

Table 2.2   Make-up of multidisciplinary subsurface team.

Job title	Job description
Subsurface manager	Manages and coordinates the work of everyone in the subsurface team.
Production geologist	Responsible for understanding and modeling the geological framework of the reservoir. Helps to identify and plan new well locations.
Geophysicist	Spends much of their time interpreting seismic data to define the reservoir structure and fault distribution. Where the seismic data allow, depositional environments and rock and fluid properties can also be characterized.
Petrophysicist	A key task is to analyze wireline logs to quantify the rock and fluid properties of the reservoir at the well scale.
Technical assistant	Provides technical support to the team. This includes data management, data preparation, and computer mapping.
Reservoir engineer	Predicts how much oil and gas a field is likely to produce, and may use a computer simulation of reservoir performance to analyze how the field will behave as well as taking a lead in reservoir management activities.
Production engineer	Responsible for optimizing all the mechanical aspects of hydrocarbon production from the wellbore to the surface facilities.
Production chemist	Analyses and treats problems related to scale formation, metal corrosion, drilling fluids, wax formation, and solids precipitation between the reservoir and the surface facilities.
Drilling engineer (well engineer)	Plans the mechanical aspects of any well operations including drilling new wells.
Economist	Costs and evaluates any economic activity relevant to the subsurface.

Limits of Reservoir Modelling", one of which described a process of what is now termed "ensemble modelling", this was the Big Loop™ workflow from Emerson. The main thrust behind the workflow and the software designed to facilitate it is that the individual technical disciplines must

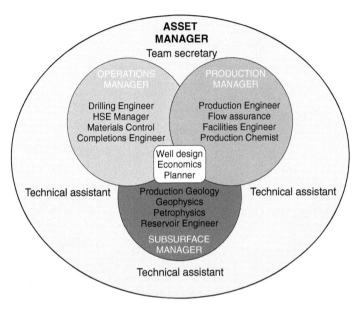

Figure 2.4   Ideal arrangement and interactions of a multidisciplinary asset team managed by different disciplines during the life of the project.

collaborate throughout the life of a field, not just co-operate with each other. Collaboration is about connecting the specialists to ensure that they listen to one another, take an interest in what the other team members are doing, bring their own expertise to a problem, and have a stake in providing a comprehensive solution to an overall problem. To achieve success with such a transformation, managers must encourage their asset teams to communicate on multiple levels.

Big Loop, which is a registered trademark of Emerson, is offered as a system for integrating and automating G&G workflows with reservoir engineering, to improve reservoir risk assessment and field planning optimization (Figure 2.5). All the reservoir modelling processes are automated and can be run at will. New data are integrated and propagated forward from geophysics and petrophysics through static and dynamic modelling. Production data captured all along the field's life cycle are assimilated using a machine learning-based assisted history match solution. Not only the dynamic models, but also the static reservoir models as well as any geophysics or petrophysics processes automated in the

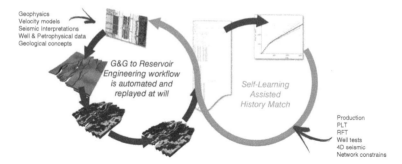

Geophysics
Velocity models
Seismic Interpretations
Well & Petrophysical data
Geological concepts

G&G to Reservoir Engineering workflow is automated and replayed at will

Self-Learning Assisted History Match

Production
PLT
RFT
Well tests
4D seismic
Network constrains

Figure 2.5 "Big Loop": automated workflow allied with machine learning-based assisted history match provides geologically consistent models calibrated to all available production data. *Source:* Illustration from Emerson ©

integrated workflow are calibrated by field data – this is the feedback part of the loop. It provides evergreen models and robust risk quantification through the generation of multiple models to capture and propagate uncertainties.

2.5 Summary

The process of reservoir management is probably different for every field, but they all share the same basic components and similar objectives: to maximize the fiscal return on investment at every stage of the process. This is best achieved with a robust and well-managed technical team that will develop an integrated way of working over the life of each stage and the project as a whole.

3

Reservoir Description

Characterizing the reservoir can mean many things: to a geophysicist it means using seismic attribute data to identify structural compartments, porosity and fluid distribution, and ultimately, reservoir heterogeneity; to a geologist it is the stratigraphic layers, distribution of facies, and reservoir properties; to a petrophysicist it is the identification of flow units based on reservoir properties and fluids, with the recognition of separate compartments; and to a reservoir engineer it is the distribution of reservoir pressure, porosity, permeability, compartments, and well productivity. In reality a complete description of the reservoir requires all of these aspects and input from all of these subsurface disciplines; this is why an integrated team approach to reservoir characterization is beneficial if not essential.

To characterize a reservoir we can look at the key elements of the reservoir description in terms of the structure, stratigraphy, facies, properties of the fluids, and pressures: do this and you will have covered all the elements of an *integrated model* of the reservoir, which we will consider in Chapter 4. We get our hard data for characterizing the reservoir from the wells and seismic interpretation, but this can be enhanced with analog data from nearby fields, regional studies, and published information on similar depositional environments. Fieldwork studies are also a great way of showing non-geologists the challenges associated with modeling facies and properties at different scales of investigation.

3.1 Multi-scale Data

A typical hydrocarbon reservoir may have a volume in the order of billions of meters cubed (~10^9 m^3), whereas a core or a wireline log taken through that reservoir only represents probably 15–25 m^3, or an order of magnitude difference of ~10^{-13}. Samples from the core represent a further difference in scale until at the pore scale (~10^{-12} m^3) the order of magnitude difference between a pore and the reservoir is ~10^{-21}. The dimensions at the well test and seismic scales are obviously not as great as the pore or core data but still amount to significant under representations of the reservoir volume (Figure 3.1).

Seismic data is the primary source for horizon and fault input to the reservoir model. While we like to think that the interpreter has correctly picked the top reservoir horizon, this is often not possible due to the nature of the seismic response, and often a near top structural surface is provided. Faults data is also subject to picking difficulties as the seismic

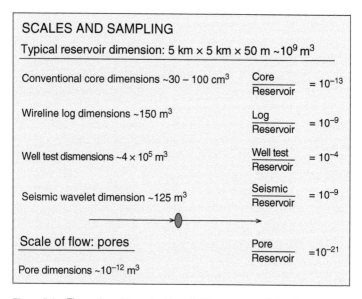

Figure 3.1   The scales of investigation of different types of data found in a reservoir study compared with the overall field size. *Source:* Cannon. S.J.C. (2018) Reservoir Modelling: a practical guide. John Wiley & Sons, Ltd. © 2018, John Wiley & Sons.

quality often deteriorates toward a discontinuity. We will look at the challenges of interpretation and depth conversion later(Section 3.2), but issues abound with respect to quality and resolution of seismic.

Well data gives the modeler the high-resolution data lacking in seismic, but because of the limited number of wells and the cost of data acquisition we generally have incomplete sets of data. Exploration wells are often not cored and may only have basic wireline logs, while production wells may have measurement-while-drilling/logging-while-drilling (MWD/LWD) data only. Well-planned appraisal or development wells will command the most comprehensive suites of logs and cores and provide the high-quality post-well analysis needed for detailed geological and petrophysical interpretation. Mud logs, cores, and core analysis data are required to calibrate the wireline information. Always remember that one of the greatest uncertainties in well data is the measurement of depth (Figure 3.2).

Wireline logs and MWD/LWD logs provide a continuous, high-resolution record of well penetration, including non-reservoir intervals, essential for depth conversion of seismic horizons and for unraveling large-scale geological sequences. Information from Dipmeter and image logs aids in structural interpretation and sedimentological analysis of sequences. Other wireline measurements that should be routinely made are reservoir pressure and fluid sampling. Comprehensive pressure gradient information is used to delineate different reservoir layers and fluid distribution and leads to a better understanding of sand body connectivity (Table 3.1).

3.2 Reservoir Structure

Hydrocarbons are trapped in many different ways to form a viable oil or gas field. Structural traps form where strata have been folded into a dome-like profile or where reservoir rocks are faulted against non-reservoir rocks, or a combination of both forms. Stratigraphic traps occur when a reservoir package abuts a non-reservoir package due to an unconformity, facies change, or where a depositional event is preserved (Figure 3.3). Identifying either type of trap is generally a function of seismic interpretation supported by well penetration. The gross rock volume of a trap is generally the least well-defined aspect of determining initial volume of

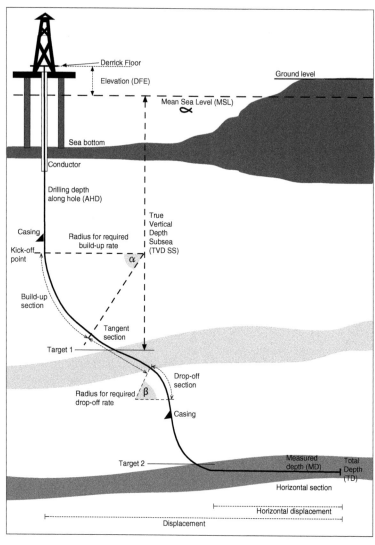

Figure 3.2 Depth measurement and well path trajectory terminology. *Source:* Cannon. S.J.C. (2018) Reservoir Modelling: a practical guide. John Wiley & Sons, Ltd. © 2018, John Wiley & Sons.

hydrocarbons that might be present. The vertical resolution of seismic can be anywhere between 10 m and 50 m depending on the quality of the data and the complexity of the overburden. Compare this to the vertical resolution of wireline log data at 15 cm and petrographic data at <1 μm.

Table 3.1 Data sources for modeling.

Static data	Application
Seismic (2D and/or 3D)	• Interpreted horizons, discontinuities, and faults • Acquisition parameters, processing
Vertical seismic profile check-shot data	• Velocity model, seismic calibration • Synthetic seismograms, time/depth calibration
Velocity calibration	• Depth conversion
Wells	• Co-ordinates, depth control, deviation, drilling history • Well pattern and development history, well stock
Cores and core analysis	• Sedimentology, petrography, environment of deposition • Porosity, permeability, grain density, fluids, shows • Petrophysical core-to-log calibration, depth shift
Mudlog	• Bulk lithology, hydrocarbon shows and kicks • Drilling history and parameters, mud losses, pore pressure
Wireline logs	• Petrophysical interpretation, porosity, water saturation • Log character and depositional environment
Dipmeter/image logs	• Well correlation, reservoir zonation, sequence stratigraphy • Structural coherence, faults, fractures • Sedimentological interpretation
Conceptual model	• Structural interpretation, depositional environment, diagenetic alteration
Seismic attributes	• Inversion, AVO anomalies, geo-body identification
Dynamic data	
Well test PLT RFT	• Inflow potential, flow rates, permeability, barriers, reservoir pressure, fluid recovery • Producing intervals, pressure gradients, and fluid contacts
Production history	• Reservoir behavior, material balance, production profiles
Fluid samples	• PVT analysis, formation water sample

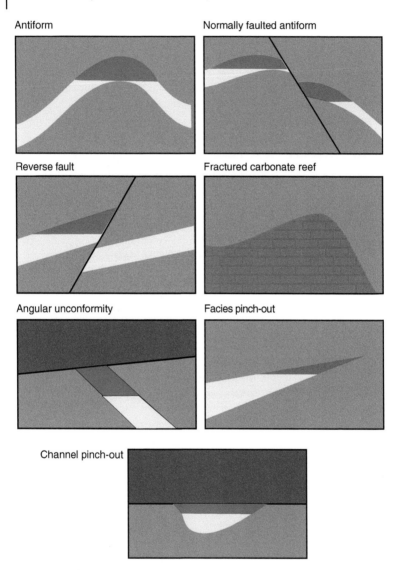

Figure 3.3 Examples of simple structural and stratigraphic trapping mechanisms that can be reproduced in reservoir models. *Source:* Cannon. S.J.C. (2018) Reservoir Modelling: a practical guide. John Wiley & Sons, Ltd. © 2018, John Wiley & Sons.

Seismic data is measured in two-way travel time: the time it takes for an acoustic wave to travel to a reflector and return to the receiver. A seismic reflection is a measure of the contrast in elasticity between two different rocks: also known as the reflection coefficient. The reflection coefficient can be estimated from the velocity (Vp) and bulk density (ρ_b) of the rocks using values from the sonic and density logs respectively (Figure 3.4). The contrast will be greatest between a well-cemented sandstone or limestone and a soft shale: the contrast between a shale seal and sandstone reservoir may not be an obvious booming event, especially where there is a gradational boundary; in other words, picking the top reservoir is not easy! It is important therefore to ask the interpreter their opinion on data quality and resolution, especially around the flanks of the structure and interpreted faults. As a rule of thumb, the vertical resolution of seismic data is between 25 and 50 m depending on the depth of interest and the overburden being investigated. Picking a fault on seismic is subject to an error of 100–200 m laterally across a surface; this could be the same as the lateral cell dimension in the dynamic model. The quality of seismic data and the area over which an interpretation is required for reservoir modeling purposes should be specified in the project work scope.

The structural model of the reservoir is built from the depth-converted seismic horizons and fault data thereby generating a reservoir framework. This is combined with the internal reservoir layering that incorporates the stratigraphic component of the model. The structural model is often designed with a gross tectonic interpretation in place reflecting the interpreter's understanding of the regional structural history; in extensional basins normal faulting is expected, whereas in compressive settings reverse faulting and slumping might be predicted. Understanding the gross depositional environment of the interval drives the internal layering of the reservoir zones, leading to stratigraphic correlation and hierarchy. We build reservoir models with non-vertical faults to improve the volumetric estimate and to better understand the distribution of hydrocarbons in the field. The modeling process can also be used to introduce separate compartments into the model, thus honoring pressure variations across the field. A fault model also aids in understanding reservoir connectivity in different parts of the field thereby improving the stratigraphic model and the understanding of trapping potential.

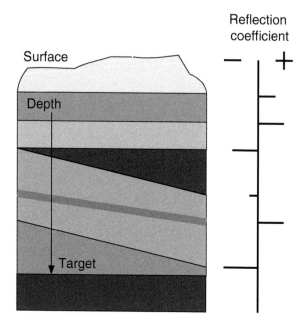

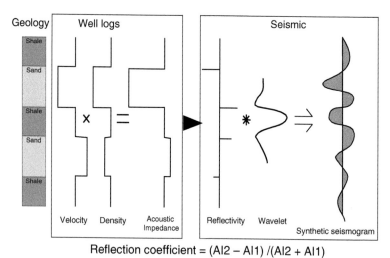

Reflection coefficient = (AI2 − AI1) /(AI2 + AI1)

Figure 3.4   Representation of the seismic response at lithological interfaces producing a reflection coefficient that may be used to generate a synthetic seismogram used in depth conversion. *Source:* Cannon. S.J.C. (2018) Reservoir Modelling: a practical guide. John Wiley & Sons, Ltd. © 2018, John Wiley & Sons.

There are a number of uncertainties in seismic interpretation, and while this is often a function of data quality, it may also be due to interpretation error. Maps defining uncertainty in the interpretations may be created to define residual maps. The main sources of seismic uncertainty are:

- *Seismic well-tie*: The reflection recorded on the seismic section may not tie the same event in a well; poor seismic resolution can result in 10–20 m mismatch. When checking the difference between a depth-converted surface and the corresponding well top ~5 m is acceptable, but more than 10 m indicates a more serious problem.
- *Seismic pick*: The reflector that is being traced may lose coherency or continuity introducing another uncertainty. Interpreters often say they could follow a reflector up or down and after depth conversion may need to make a correction to the interpretation.
- *Imaging*: The seismic response weakens with depth or below high-velocity layers as the acoustic energy is depleted, resulting in poorer amplitude response and lower resolution; around faults the acoustic waves are dissipated leading to further deterioration of the image.
- *Depth conversion*: In many ways the process that leads to greatest systematic uncertainty in seismic interpretation. For the purposes of structural modeling we will focus on this process and the uncertainties.

The experienced interpreter will appreciate these challenges exist and that a 'zone of uncertainty' exists around an interpreted horizon. This can be described as 'measurement' uncertainty as opposed to purely interpretation uncertainty. Thus, a description of the variability or tolerance can be quantified for future uncertainty workflows (Bacon et al. 2003).

Uncertainty in the reservoir structure comes entirely from the seismic interpretation and subsequent depth conversion: horizon picking, the well-tie, and velocity model introduce errors into the reservoir framework. The impact of a poor time-to-depth model can introduce large errors into the gross rock volume estimate, up to 30% in my experience. Such large errors are likely to be systematic rather than random and reflect an error in data analysis, such as using only development wells to model the overburden (Figure 3.5). Smaller errors are likely to reflect horizon miss-picks in areas of poor seismic imaging. Handling structural uncertainty in the model build will be covered in Section 4.7, Model Analysis and Uncertainty,.

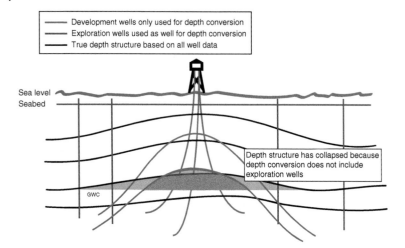

Figure 3.5 The impact of only using development wells in depth conversion; the structure collapses because the depth conversion is based on a limited overburden distribution and does not include offset wells. In this case the in-place volume was reduced by approximately 30%. *Source:* Cannon. S.J.C. (2018) Reservoir Modelling: a practical guide. John Wiley & Sons, Ltd. © 2018, John Wiley & Sons.

3.3 Reservoir Framework

The reservoir structure is completed by the addition of stratigraphic levels represented by seismically interpreted horizons/events and geologically significant surfaces identified in well data: where the levels are identified in both datasets then the mapped seismic horizons are constrained by the well picks. All that is needed to build the reservoir framework is a top reservoir horizon and a base, ideally derived from seismic interpretation. Internal stratigraphic levels are usually calculated from correlatable horizons seen in the well data; these are often incorporated as isochore thickness maps. The internal zonation should reflect major changes in geology that have some influence on flow in the reservoir. The changes could be in the gross depositional environment, or in the type or style of heterogeneity or a change in the facies type or orientation.

Well correlation provides a fundamental constraint on the reservoir framework, so as it is a deterministic element of the model you cannot afford to get it wrong. The key workflow activity is defining those

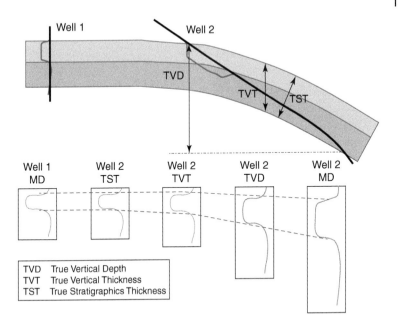

Figure 3.6 Depth reference terminology used in well correlation. *Source:* Cannon. S.J.C. (2018) Reservoir Modelling: a practical guide. John Wiley & Sons, Ltd. © 2018, John Wiley & Sons.

surfaces that can be identified across the field in *every* well that is to be used in the model, unless there are erosive events recognized. The surfaces may be chronostratigraphic, lithostratigraphic, or based on pressure data; they should not be just 2D correlations based on petrophysical layer average data. Well correlation should also be done using true stratigraphic thickness) (TST) as the reference, because true vertical depth (TVD) will stretch/squeeze the thickness, especially of highly deviated wells (Figure 3.6).

Well correlation leads to zonal attribution in the model, so should be kept simple and reflect the conceptual geological model and explain the observed differences in thickness of a zone between wells. After picking a specific marker in a well, a depth map should be generated; the map will reveal any anomalous picks, allowing the modeler to review the well pick with the geologist. Isochore maps should also be generated to look for trends in the data or unexpected thickness changes. Isochore maps become a key construction element in building the detailed reservoir framework where there is sufficient well data to use geostatistical mapping tools to create the map.

3.4 Depositional Environment

Most hydrocarbon reservoirs are found in either clastic or carbonate environments; on the basis of the global distribution of oil, carbonate reservoirs are more important, but much more difficult to develop efficiently. Based on the large-scale wireline log shapes and the associated cores it is possible to build up a picture of the environment of deposition of the reservoir interval and any significant non-reservoir packages. Most reservoirs can be classified into one or more contiguous depositional environments with the commonest being aeolian, lacustrine, fluvial,

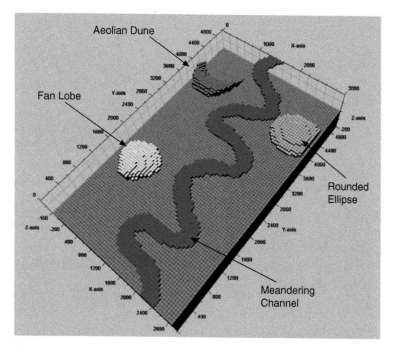

Figure 3.7 Examples of different types of genetic units that can be modeled using an object modeling method. *Source:* Cannon. S.J.C. (2018) Reservoir Modelling: a practical guide. John Wiley & Sons, Ltd. © 2018, John Wiley & Sons.

deltaic, shallow marine, and deep marine; there are many other combinations of reservoir types, but these extend beyond the scope of this book.

The large-scale architecture of reservoir and non-reservoir units is a major control on hydrocarbons initially in-place (HIIP), drainage, and sweep efficiency. This has an impact on the volume of hydrocarbons that can be recovered. Drainage and sweep efficiency are governed by the connectivity of the reservoir units. This is primarily governed by the net: gross. The identification of genetic reservoir units (such as channels, bars, etc.) is a key step in modeling the large-scale reservoir architecture (Weber and van Geuns 1990). These units form the basis of object-based reservoir modeling techniques (Figure 3.7). The impact of large-scale reservoir architecture on flow depends not only upon the nature of the architecture, but also upon the fluid properties and flow regime. The impact of heterogeneity on flow is accentuated if the mobility ratio is unfavorable.

3.5   Static Reservoir Properties

The petrophysical properties of a reservoir are needed to estimate the storage capacity and flow potential and ultimately the distribution of water, oil, and gas in the structure. Essentially, we are trying to estimate the porosity, hydrocarbon saturation, and permeability of a given rock type. The measurements we make in the subsurface are based on the density and resistivity of the rocks, as well as their acoustic and nuclear properties, taking into consideration the fluids they contain. None of the measurements are direct; they are all indirect values based on interpretation of a physical response to a static property or some source of stimulation. The measurements must be calibrated by core analysis data, which is a direct, if imperfect, measurement, before they can be used for estimation of the desired property we hope to model.

Porosity (*Phi*) is defined as the capacity of a rock to store fluids and estimated as the ratio of the pore volume to the bulk volume. Porosity is a nondimensional parameter expressed as a fraction or percentage (Figure 3.8a). The porosity of a rock comprises two main elements: primary depositional or intergranular porosity, and secondary porosity, which may be the result of grain or particle dissolution or present as microporosity in authigenic clays. Porosity may be defined as *effective* or *total* depending on whether it includes porosity associated with clays; some tools measure total porosity and must be corrected for the clay content. This is a simple classification that does not include all carbonate rocks or certain clay-rich shale reservoirs. Fractured reservoirs need also to be treated separately, being defined as having a dual porosity system: matrix and fracture.

Water saturation (S_w) is the proportion of total pore volume occupied by formation water; hydrocarbon saturation is derived from this relationship ($S_h = 1-S_w$). It may be expressed as a fraction or a percentage depending on how porosity is defined (Figure 3.8b). There is another direct link to porosity terminology as water saturation can be either a total or effective value. Logs measure both the mobile water and the clay-bound water in the pore space. The terms irreducible, residual, connate, and initial water saturation are also commonly used, sometimes without due regard to the meaning. Irreducible water saturation (S_{wirr}) is defined as the minimum S_w at high capillary pressure and saturation, as the effective permeability to water approaches zero. The initial water saturation (S_{wi}) is the proportion of water in the reservoir at the time of discovery and may be synonymous with connate water, the water saturation at time of deposition, if no hydrocarbons are present. In a hydrocarbon-bearing reservoir S_{wirr} is always less than S_{wi}. The term 'transition zone' also has more than one meaning depending on who is using it: to a geologist or petrophysicist it is the zone between the lowest level of irreducible water and the free water level; this is a static definition. To a reservoir engineer it is an interval in a well that flows both oil or gas and water at the same time: the two 'zones' may be contiguous.

Permeability (K or k) is the measure of the capacity of a reservoir to conduct fluids or for flow to take place between the reservoir and a wellbore. Although a dynamic property, permeability is dependent on the associated rock and fluid properties (Figure 3.8c); it is also one of the most difficult to measure and evaluate without data at all relevant scales – core, log, and production test. At the microscopic or plug scale, permeability is a function of pore network and whether there are large or small pore throats and whether the connecting pathways are straight or tortuous, a function of grain size and sorting. Permeability is also a vector property as it may have a directional component, resulting in anisotropy. Permeability may vary greatly between the horizontal and vertical directions, impacting on the

(a)

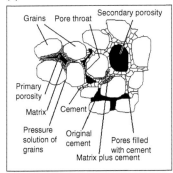

Porosity = $\dfrac{\text{volume of pore space}}{\text{total volume of rock}}$

Expressed as a fraction or percentage
Afunction of grain size and packing
Can be expressed as TOTAL or EFFECTIVE

Primary porosity reduces with compaction due to burial and lithification/cementation
Secondary porosity is a result of dissolution of unstable minerals.

(b)

Water saturation (S_w) = $\dfrac{\text{pore volume filled with water}}{\text{total pore volume}}$

Hydrocarbon saturation (S_h) = (1-S_w)
Can be expressed as TOTAL or EFFECTIVE property dependent
Bec onsistent in use of terms

(c)

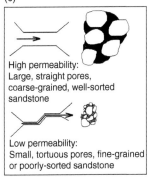

High permeability:
Large, straight pores, coarse-grained, well-sorted sandstone

Low permeability:
Small, tortuous pores, fine-grained or poorly-sorted sandstone

Measure of the ability of a reservoir to conduct fluids
A dynamic property dependent on rock and fluid characteristics with a directional (vector) component
May be a predictable relationship with porosity

$$Q = \frac{kA\Delta p}{\mu L}$$

Absolute, effective, and relative permeability values may be required as a key input for dynamic simulation

Figure 3.8   (a) Porosity: the relationship between volume of pore space and total volume of rock is a function of grain size, sorting, and packing at time of deposition. Post-depositional processes such as compaction and diagenesis can alter the original relationship. (b) Water saturation: the proportion of the total reservoir pore volume filled with water: the remaining pore volume is filled with oil or gas, not necessarily hydrocarbon gas. (c) Permeability: the ability of a reservoir to conduct fluids through an interconnected pore network. *Source:* Cannon. S.J.C. (2018) Reservoir Modelling: a practical guide. John Wiley & Sons, Ltd. © 2018, John Wiley & Sons.

directional flow capacity of a reservoir. Given the difficulties in reliably measuring permeability, a qualitative assessment is often made depending on the hydrocarbon in-place (Table 3.2).

Permeability is measured in darcies (D) but usually reported as millidarcies (mD), named after the French water engineer who first attempted to measure the flow of water through a vertical pipe packed with sand. The rate of flow is a function of the area and length of the pipe, the viscosity of the fluid, and the pressure differential between the ends of the pipe. This law only applies to a single fluid phase and may be termed absolute or intrinsic permeability. Effective permeability (K_{eff}) is the permeability of one liquid phase to flow in the presence of another; relative permeability (K_r) is the ratio of effective to absolute permeability for a given saturation of the flowing liquid, that is, permeability of oil in the presence of water (K_{ro}). Permeability is a key input for numerical reservoir simulation.

Relative permeability is the normalized value of effective permeability for a fluid to the absolute permeability of the rock. Relative permeability expresses the relative contribution of each liquid phase to the total flow capacity of the rock.

$$k_{ro} = k_o / k, k_{rw} = k_w / k, k_{rg} = k_g / k$$

where k_o, k_w, and k_g are the effective permeability to each potential fluid phase. The measurement of relative permeability is fraught with difficulties and results must be treated with care. Wettability issues as well as small-scale heterogeneities in the sample affect measurements; consideration must be given to these and other experimental issues when evaluating the results for use in dynamic simulation.

Table 3.2   Permeability ranges for different qualitative descriptions of permeability.

Poor	<1 mD	'tight' for gas
Fair	1–10 mD	'tight' for oil
Moderate	10–50 mD	
Good	50–250 mD	
Excellent	>250 mD	

Examination of typical relative permeability curves (Figure 3.9), reveals that the relationship between the relative permeability and saturating phase is nonlinear. For the oil phase, a relative permeability value of 0 is encountered at the limiting (end point) saturation where oil ceases to flow in the reservoir: this is known as the residual oil saturation. In the case of the water phase, the end point saturation where water is immobile is referred to as the irreducible water saturation and can be related to the entry pressure of a capillary pressure experiment. Values of k_{ro} and k_{rw} between 0 and 1 indicate simultaneous flow of oil and water: this only occurs in a transition zone. When $k_{ro} = 1$, only the oil phase is flowing; when $k_{rw} = 1$, only the water phase is flowing.

Relative permeability is measured either in a steady-state or unsteady-state experiment. In a steady-state experiment a fixed ratio of liquids is flowed through the sample until pressure and saturation equilibrium is reached; achieving steady-state flow can be time consuming, especially

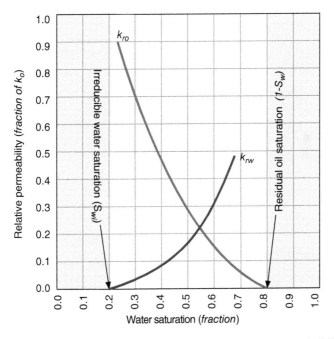

Figure 3.9 Relative permeability of a rock sample to water and oil liquid phases: this is two-phase relative permeability where $k_{ro} = 0$, $S_{or} = (1-S_w)$.

in less permeable material. The effective permeability of each liquid is calculated as a function of the relative saturation using Darcy's law, by measuring the flow rate, pressure differential, and saturation. Monitoring the total effluent from a core sample during an imposed flood and calculating the relative permeability ratio that is consistent with that outcome is the basis of unsteady-state measurements. Steady-state experiments are more reliable and accurate but take longer than the cheaper unsteady-state tests, which provide a greater interpretational challenge.

Traditionally, core porosity is plotted against the logarithm of core permeability to give a linear poro-perm relationship, beloved of geologists and petrophysicists. This linear relationship is then used to evaluate reservoir quality and to distribute permeability in a model. The relationship is at best tentative as permeability is function of grainsize and sorting, which define the pore throat size distribution. Wherever possible the facies model should be used to partition the core data such that a number of poro-perm relationships can be derived that will be used to define heterogeneity.

Capillary pressure acts at a microscopic scale in the reservoir, which in conjunction with viscous and gravitational forces define how a reservoir performs dynamically. Capillary pressure occurs whenever two immiscible fluids occur in the pore space of a rock and is defined as the pressure difference measurable in the two phases (Figure 3.10a). There is an inherent relationship between capillary pressure and water saturation because water is retained in the pore space by capillary forces. Capillary pressure also determines the fluid distribution and saturation in a reservoir, hence the link to wettability.

Wettability is a measure of a rock's propensity to adsorb water or oil molecules onto its surface in the presence of the other immiscible fluid. At deposition, a thin film of water is usually formed around the grains leaving the rock water wet, the normal situation; however, carbonate rocks are commonly oil wet or have intermediate wettability. Wettability is a function of the surface tension between the solid grain and the fluid in the pores (Figure 3.10b).

The impact of wettability on the other dynamic properties of a rock is important to understand as it controls the fluid saturation and distribution in a reservoir. While most (clastic) reservoirs would be

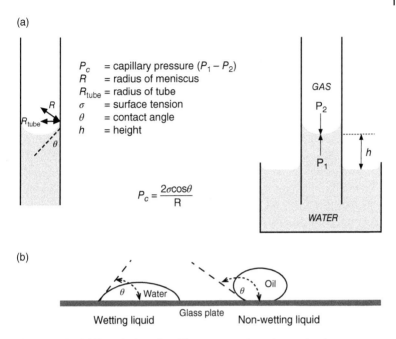

Figure 3.10 (a) Description of capillary pressure based on a simple experiment of water rising in a tube. (b) Description of wettability as the interaction between a surface and an adsorbed fluid. *Source:* Cannon. S.J.C. (2018) Reservoir Modelling: a practical guide. John Wiley & Sons, Ltd. © 2018, John Wiley & Sons.

considered to be water wet, under certain conditions all reservoirs can become oil wet at least in part. Carbonate reservoirs have a greater tendency for the oil wet state because of the greater adsorption capacity of calcium/magnesium carbonate. Many reservoirs are of mixed wettability: oil wet in the large open pores and water wet in the smaller isolated pores often filled with microporous clays.

3.6 Dynamic Reservoir Properties

A good starting point for a discussion of the dynamic properties of a reservoir is to look at the fluid data. The fluids present in a hydrocarbon reservoir are water, oil, and/or gas, and given their different physical properties they are generally segregated in the reservoir column. The distribution of these fluids provides information on the vertical and lateral compartmentalization of a reservoir. The composition of the fluids both in terms of chemistry and pressure–volume–temperature (PVT) properties may further indicate heterogeneity in the reservoir. These variations represent changes in the fluid distribution over geological time since the reservoir was initially filled by formation water and subsequent migration of hydrocarbons.

Almost all hydrocarbon reservoirs are deposited in an aqueous environment or are soon after filled with water of varying degrees of salinity: a freshwater aquifer or a marine formation water. The tendency is for formation water to become increasingly saline over time due to burial, compaction, and chemical concentration of dissolved minerals. Where reservoirs are more deeply buried a variation to this rule of thumb has been observed by the author, such that water samples from the hydrocarbon leg are more saline than the water leg: one explanation is that below ~3000 m a chemical change from smectite clays to illite clays releases water of crystallization which is effectively distilled water, thus diluting the irreducible water trapped in the pore space.

3.6.1 Pressure Measurements, Gradients, and Contacts

There are three basic descriptions of pressure in a borehole: hydrostatic, overburden, and formation pressure (Figure 3.11).

- *Hydrostatic* pressures are those due to a connected fluid column from the surface to a given depth on the subsurface; the hydrostatic gradient is a function of depth and fluid density and is also known as 'normal pressure'.
- *Overburden* pressure is the total pressure exerted by the weight of the overlying rock and the pressure of the formation fluids; the overburden gradient is a function of the bulk density and height of the rock column. This is also known as the principle stress direction in geomechanical terms.
- *Formation* pressure is the pressure of the fluids contained in the pore spaces of the sediments and can be 'normal', 'subnormal', or 'abnormal' also termed underpressured or overpressured respectively; both may be hazardous during drilling.

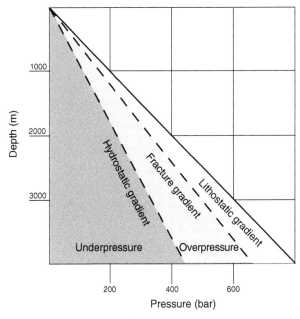

Calculations and conversions:

Pressure (Psi) = MWppg*0.0519*TVDft

Pressure (Bar) = MWg/cc*0.0981*TVDm

Figure 3.11   The pressure gradients that are to be expected in the subsurface as a well is drilled. The formation pressure generally lies between the hydrostatic and lithostatic gradients unless the strata is overpressured. *Source:* Cannon. S.J.C. (2016) Petrophysics: a practical guide. John Wiley & Sons, Ltd. © John Wiley & Sons.

For all three calculations, estimates of the fluid density and bulk density are required, and wireline measurements provide both of these variables either directly or indirectly. Other data can be gathered during drilling to calibrate these formation parameters, such as cuttings, gas measurements, and LWD logs. All pressures must be recorded in consistent units, either bar per meter (b/m_a or b/m_g) or pounds per square inch (psi_a or psi_g): where 'a' stands for atmospheric and 'g' stands for gauge. This simple datum variation of 14.7 psi has been the cause of numerous mistakes in the integration of data.

3.6.2 Fluid Contacts

The oil–water, gas–oil, or gas–water contact in a reservoir is a surface of hydraulic equilibrium across a field, unless the existence of different faulted compartments breaks up this equilibrium. It is worth noting that just because wells have the same fluid contact, they may not be in hydrodynamic communication as a result of diagenetic alteration near faults. In this case, it is not until after production starts that variations will be observed in the rise in water level due to fluid withdrawal. Often, sedimentological heterogeneity, recorded as a change in facies or rock type, may result in a hydrocarbon down-to or water up-to depth measurement. The strange phenomenon of a tilted contact is generally associated with hydrodynamic activity in the aquifer. While each of these contacts might be observed in well data the only physically significant surface is the free water level: the surface at which the capillary pressure in the reservoir is zero and the water saturation is 100%.

Wireline log data and pressure measurements are the most common tools used to locate fluid contacts. The ubiquitous wireline formation tester is one of the most effective ways to identify the hydrocarbon–water contact in appraisal and development wells. These tools basically insert a metal probe into the borehole wall to measure the static pressure, induce a pressure draw down, and measure the response of the pressure buildup; they are also designed to collect single or multiple formation fluid samples. From the pressure buildup and also the ease of sampling an estimate of permeability can be made. The measurements are subject to many ambiguities, especially in low permeability zones or where the borehole is badly washed out. Using the results of wireline formation tester (WFT) measurements requires care and attention to the basic data acquisition parameters; depth, pressure datum, environmental conditions, and the accuracy or precision of the gauges used.

Having established a reliable and consistent set of pressure data it can be used to investigate fluid gradients in single or multiple wells through simple graphing methods (Figure 3.12); always use the TVD measurement to compare wells. Having plotted the pressure data correctly against depth, one or more trends may be identified that will relate to gas, oil, or water gradients. Slopes with a gradient of 0.0233–$0.032\,\mathrm{b\,m^{-1}}$ ($<0.31\,\mathrm{psi\,ft^{-1}}$) are likely to represent gas or gas condensate; 0.069–$0.087\,\mathrm{b\,m^{-1}}$ (0.32–$0.36\,\mathrm{psi\,ft^{-1}}$) light oil, 0.53–$0.58\,\mathrm{b\,m^{-1}}$ (0.37–$0.41\,\mathrm{psi\,ft^{-1}}$) heavier oil, and $0.61\,\mathrm{b\,m^{-1}}$

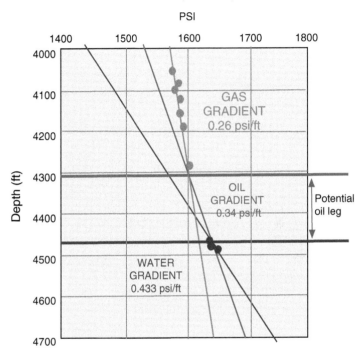

Figure 3.12 Typical oil, gas, and water gradients displayed against depth and pressure revealing a potential oil leg.

$(0.433\,\text{psi}\,\text{ft}^{-1})$ is the gradient for freshwater. Where these gradients intersect identifies the fluid contact under investigation (Table 3.3).

These gradients can be related back to the specific gravity of the liquid through the equation:

$$\text{Gradient: psi}\,\text{ft}^{-1} = \text{specific gravity} / 2.31 = 0.09806\ \text{b m}^{-1}$$

where specific gravity is in $\text{g}\,\text{cm}^{-3}$. The relationship between specific gravity and oil density in API degrees is given by:

$$\text{Oil density at surface conditions: API} = \left(141.5 / \text{specific gravity}\right) - 131.5$$

Table 3.3 Typical pressure gradients and fluid densities.

Fluid	Gradient bar (m^{-1})	Gradient (psi ft^{-1})	Density (g cm^{-3})
Dry gas	0.022	0.100	0.230
Wet gas	0.032	0.140	0.320
Oil limit	0.069	0.300	0.689
Light oil 60 °API	0.087	0.387	0.780
Heavy oil 20 °API	0.091	0.404	0.934
Freshwater	0.098	0.433	1.00
Sea-water	0.101	0.444	1.025

As the gas–oil ratio of the liquid increases the apparent density of the oil at reservoir conditions will be reduced leading to a lowering of the repeat formation tester (RFT) gradient.

Gradients in the aquifer can be used to estimate formation water salinity if R_w and bottomhole temperature (BHT) are known, using the standard conversion chart. For example:

A water gradient of 0.45 psi ft^{-1} is equivalent to a specific gravity of 1.04 g cm^{-3}, which is translated as a salinity of 60 000 ppm NaCl at 200 °F (93 °C) or a resistivity of 0.046 Ω.m.

3.6.3 Well Test Data

Well tests have been around since drilling began; even a surface blowout beloved of drillers of onshore wells more than 100 years ago gives an indication of how a reservoir may perform. Today, with the advent of high accuracy electronic gauges and computer software, much more can be done to collect data and interpret the results. Well test analysis using type curves is still highly interpretive and the results are nonunique. The modern quartz gauges in use today allow for a variety of special well tests such as pulse tests and interference tests between wells to estimate reservoir connectivity. Many fields also have permanent downhole gauges installed to provide continuous pressure profiles during production. During appraisal and development, well test analysis can reveal whether there are boundaries or heterogeneity near to a well and thus confirm the presence of a sealing fault of facies change.

3.6.4 Transient Well Tests

Pressure buildup and drawdown tests are routinely used to estimate flow rates, production index (PI), well bore skin, and permeability. Although 'transience' or 'infinite acting behavior' is an assumption in most analytical solutions to these data, 'it should never be assumed but must always be proven' (Dake 2001). For a complete well test analysis, it is essential that an understanding of the interval being tested is made, with an understanding of heterogeneity of the interval pre-eminent.

The commonest type of well test is the pressure buildup test, where, ideally, the well is produced at a constant rate (q stb d^{-1}), for a fixed time (t hours), after which it is shut-in for a pressure buildup. Pressure measurements are recorded during both phases of the test and in the case of the buildup as a function of the closed-in time. These data, rate, pressure, and time are used to determine the reservoir pressure and formation characteristics. The validity of the well test interpretation will be dependent on the availability of core and wireline log data to establish reservoir quality, pressure–depth profiles derived from RFT measurements, a conceptual model of reservoir geology, the drive mechanism, PVT properties, the well completion equipment, and the available downhole and surface hardware, especially the type and sensitivity of the gauges.

Well tests are performed throughout the lifetime of a field, often on a regular basis, to quantify the production behavior of individual wells and to determine the average reservoir pressure around the well bore. A compromise is always required between the acquisition of data during the shut-in period and loss of production. The shut-in time will be a function of fluid mobility and reservoir quality. Not all reservoirs will achieve pressure equilibrium during the specified shut-in period and a theoretical static reservoir pressure will be estimated from the buildup phase. It is more important that consistent measurement of the pressure decline in the reservoir due to production is recorded or calculated for estimates of ultimate recovery and further planning.

3.7 Reservoir Hydrocarbon Fluids

The type of reservoir fluid is a major factor in the production behavior and exploitation of a reservoir. All reservoir engineering applications require some understanding or assumptions of the PVT behavior of the reservoir fluids. PVT properties are a link between surface facilities and reservoir conditions. These data always carry a high degree of uncertainty resulting from sampling methods and strategy, vertical and lateral variations, to post migration alteration such as biodegradation. When working with PVT data it is important to use as much experimental data as possible; any gaps in data can be field correlations or analogs or equation-of-state (EOS) calculations. Therefore, the understanding and integration of all available data helps to reduce this uncertainty.

The properties of a reservoir hydrocarbon fluid are a function of their chemical composition and reservoir pressure and temperature: these conditions determine whether the fluid in the reservoir is a gas or a liquid. The four most common hydrocarbon fluids can be described by a phase diagram (Figure 3.13) that defines the bubble point, dew point, and critical point of a given fluid type: undersaturated (crude) oil, volatile oil, gas condensate, or gas. The vertical dotted line in Figure 3.13 represents the change in pressure and temperature as a reservoir is produced: in reality, reservoir temperature does not significantly change, but only at the surface is there an impact of the fluid type.

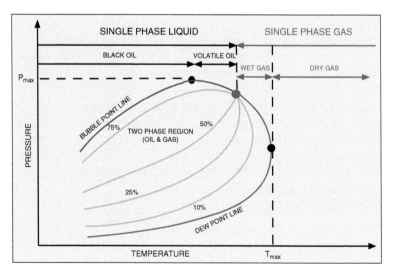

Figure 3.13   Standard phase diagram depicting liquid and gas phases of different hydrocarbon types.

By following the vertical line, it is possible to see the change in fluid character with change in reservoir pressure. The crude oil phase envelope, initial pressure, and temperature describe an undersaturated hydrocarbon; as the pressure drops below the bubble point line, some gas is liberated. The phase behavior of a volatile oil is similar except that significantly more gas is liberated. In the case of a gas condensate, the reservoir temperature is higher than the critical temperature; however, the critical point is to the left of the two-phase region so that as pressure is reduced the dew point line is crossed and a small amount of gas condenses forming a liquid phase. In the case of a dry gas the reservoir temperature is always greater than the critical point, even during depletion, and no liquids are produced. This is because of the initial composition being of lighter elements such as methane and ethane.

During reservoir depletion, oil flows through the production wells to the surface because the pressure at the base of the well exceeds the hydrostatic head of the column of oil in the well. Over time the oil rate decreases as the initial reservoir pressure decreases. If reservoir pressure falls below the oil bubble point pressure, gas initially dissolved in the oil comes out of solution and flows preferentially to the production well, because it has a much lower viscosity. At the same time the viscosity of the remaining oil increases, reducing its mobility further. This will further reduce the oil production rate, although the total (oil plus gas) production rate might increase through reducing the hydrostatic head in the well. Water (or gas) injection is usually applied before this happens so as to maintain reservoir pressure above the bubble point.

Oil and gas fluid data are required to evaluate the properties of produced fluids at reservoir conditions, in production tubing, in process facilities, and in pipeline transportation. The key PVT properties to be determined for a reservoir fluid include:

- Original reservoir fluid composition(s).
- Saturation pressure at reservoir temperature.
- Oil and gas compressibility.
- Oil and gas density.
- Oil and gas viscosity.
- Gas solubility in reservoir oil (gas–oil ratio [GOR], Rs).
- Liquid content of a reservoir gas.

- Shrinkage (volume) factors (B_o, B_g, B_w) of oil, gas, and water from reservoir to surface conditions.
- Compositional variation with depth.
- Equilibrium phase compositions.

Reservoir fluid volumes are generally reported in stock tank volumes, and the shrinkage factor (formation volume factor [FVF]) is therefore a critical property. It should be noted that this property is related to the actual process by which the type of reservoir fluid is established. Usually, shrinkage factors are calculated by EOS simulations. Experimental data are used indirectly, to tune the EOS parameters (Table 3.4).

The FVF expresses the degree of shrinkage a barrel of oil undergoes in the journey from the reservoir to the surface processing facility. There are three terms to remember: the B_o, B_g, and R_s:

- B_o is the ratio of the oil volume at reservoir condition to the volume at the surface; this is always greater than unity.
- B_g is the ratio between the volume of free gas at reservoir conditions and the volume at the surface: this is not the same as E, the gas expansion factor.
- R_s is the solubility ratio, which quantifies the volume of surface gas that dissolves in one stock tank barrel of oil at reservoir conditions: the gas–oil ratio.

Table 3.4 Typical properties of the major reservoir hydrocarbons.

	Crude oil	Volatile oil	Gas condensate	Wet gas	Dry gas
Initial GORscf/stb	< 2000	2000–6000	6000–20 000	20 000–100 000	> 100 000
Color	Black to light green	Dark straw	Straw to colorless	Colorless	Colorless
API gravity	10–45	40–50	45–65	n/a	n/a
Composition	$C_{7+} > 40\%$	C_{7+} 12.5 to 40%	C_{7+} 2 to 12.5%	Primarily C1-C2	Primarily C1
Oil FVF, rb/stb	< 2.0	> 2.0	n/a	n/a	n/a

The composition of hydrocarbons within a field may vary for a number of reasons, including source rock composition, timing of charge, biodegradation, and gravitational segregation. Given a sufficient amount of time, convection and diffusion in the reservoir will tend to homogenize these differences; however, where they exist may indicate compartmentalization. Many reservoirs exhibit vertical and/or lateral variations in PVT properties. Gravity segregation is a common feature in steeply dipping or very thick reservoirs. Faults often compartmentalize a field resulting in differing saturation history during migration. Laterally extensive reservoirs may exhibit various fluid types especially in fields with heavier oils. The key to understanding such variation is often in reviewing sample strategy and reliability to help decide what is real and what is an artifact.

3.8 Summary

An oil or gas reservoir is a complex entity and one that requires careful analysis. To do this careful analysis you need sufficient data collected over time and experience to know what this data is telling you, because it will almost always be contradictory. Do not be tempted to ignore data that doesn't fit your 'model' or to manipulate data so it supports one interpretation only; it will only come back to bite you later!

4

Building an Integrated Reservoir Model

To better explain how to build an integrated reservoir model I am going to start at the end of the process and then show how we got there. For the purposes of this book we want to end up with a dynamic simulation of the reservoir that we can use for field management, production control, and prediction of ultimate recovery. We also want a model that can be easily updated as data are gathered during development and production. This is not a trivial task, as any reservoir engineer would tell you!

The key to building an integrated reservoir model is not the software; it is the thought process that the reservoir modeler has to go through to represent the hydrocarbon reservoir they are working on. This starts with a conceptual model of the geology and a diagram of the 'plumbing' model to represent how fluids might flow in the reservoir (Figure 4.1). Figure 4.1 shows the vertical and lateral distribution of a series of sand bodies, together with the permeability distribution and fluid gradients encountered in one well. Modern integrated modeling software starts with seismic input in terms of both interpreted horizons and faults, and seismic attribute data that characterizes reservoir from non-reservoir and ends with a link to dynamic simulation; the so-called seismic-to-simulation solution. A few key statements should be made at the outset:

- Every field is unique and therefore has different challenges.
- Every challenge will have a unique solution.
- Every solution is only valid for the given situation and therefore. . .
- KEEP IT SIMPLE . . . at least to begin with.

Reservoir Management: A Practical Guide, First Edition. Steve Cannon.

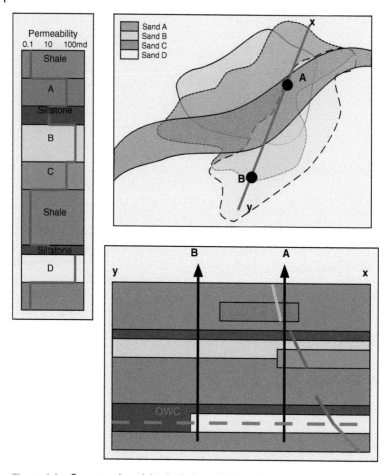

Figure 4.1 Conceptual models: depictions of the sedimentological, stratigraphic, and 'plumbing' models of a reservoir showing the permeability distribution and fluid gradients present in the different sand bodies.

Numerical simulation is the default tool for any field development problem that an engineer might encounter, notwithstanding the fact that most engineering managers will expect a simulation-based solution to nearly every question posed from well placement to ultimate recovery. Modern simulation packages have pre- and post-processing functions to maximize the use of the results of simulation. However, simulation should only be used to answer a specific, well-posed question of economic importance; this is because of the time and effort that goes into gathering the data and building the model. An adequate reservoir description must be available that can be incorporated in the solution as well as an answer that is dependent on understanding the changing dynamic properties of the reservoir during production. If possible, only build a dynamic reservoir model if alternative analytical solutions have been found to be no longer sufficient or valid, before running a simulation.

It is important that the asset team manages expectations because reservoir simulation will only provide an approximate, but repeatable, answer dependent on the quantity and quality of the input data. Uncertainty in the results might be due to an incomplete understanding of the geological model, sample bias in the available data, or an inappropriate choice of the type of numerical solution, such as grid or numerical dispersion. It is also important to realize that the results of a simulation are nonunique such that different reservoir descriptions can give the same history match of production but provide different production forecasts. Having said that, simulation is the only way to provide production profiles and cash flow predictions and, more importantly, provide production profiles under different development scenarios.

Finally, never start a simulation study without a well-defined problem and clear objectives. Determine which model parameters have the greatest influence on the study outcome and also whether you have sufficient data to resolve the issue. This is a clear case of garbage in leads to garbage out.

4.1 Simulation Model Design

Before getting into the details the following questions need to be addressed:

1) What is the reservoir drive mechanism?
2) What is the type and quality of data available?
3) What resources, hardware, and software, as well as people are available?
4) What is the question that needs an answer?

By addressing these questions, it is possible to decide on the type of model geometry, the overall area of interest, and the type of simulator required for the project.

4.1.1 Drive Mechanism

There are five types of basic natural drive mechanisms observed that relate the reservoir energy to some kind of fluid expansion process and are used to classify the dynamic behavior of the reservoir:

1) Fluid expansion
2) Solution gas drive
3) Water drive
4) Gas cap drive
5) Compaction drive

Most reservoirs show a combination of mechanisms, especially as hydrocarbon production continues, such that an undersaturated oil field that initially produces a single phase, oil, under fluid expansion until the reservoir pressure falls below bubble point, at which point gas is released and a solution gas drive mechanism takes over. Understanding the nature of the drive mechanism is essential when considering a dynamic simulation study of a field. We will look at each drive mechanism in more detail in the Chapter 5.

4.1.2 Type and Quality of Data

In Chapter 3, we looked at the type of data that are required to build an integrated reservoir model of an oil or gas field; however, all those data are seldom available nor unambiguous. We need to have a top structure map and a confirmed hydrocarbon fluid contact to correctly estimate the gross rock volume of the field and a porosity model and saturation model to estimate the pore volume and hydrocarbon volume initially in-place. We need to know the initial reservoir pressure and the fluid type, and ideally have a representative sample for pressure–volume–temperature (PVT) analysis; with this data we can start to initialize the model.

Before the advent of 3D geological modeling at the end of the twentieth century, reservoir engineers would build simulation models with numerous layers each with a layer average value for porosity, water saturation, and permeability; the water saturation might be distributed vertically using a saturation-height model. The definition of the layers was often done based on the correlation of well-derived petrophysical properties. These models were generally too simple compared with the geological understanding of the field and often resulted in a very 'smooth' dynamic response to the flow simulation. Also, these models would have vertical faults resulting in the incorrect allocation of hydrocarbon volumes in adjacent compartments. This is the simplest, direct measurement of volumetric estimation and is discussed in Chapter 5.

Today, a reservoir simulation model will be based on a detailed geological model, ideally incorporating a facies model that is used to stochastically distribute the porosity often as a rock-type model; water saturation distributed with a saturation-height model based on a well-defined free water level tied to the rock type; and a permeability distribution based on core, log, and well test data that support the multiscale nature of the property (Figure 4.2). A modern reservoir model will have a representative structural framework that includes inclined faults to correctly allocate the hydrocarbon volumes, and of course a seismically-derived, depth-converted top structure surface. Finally, by incorporating a stratigraphic framework it is possible to reduce the number of hard layers by building a finer grid of cells, each with a single value of each property to be modeled (Figure 4.3).

Reservoir modeling is a challenge because we are dealing with a mix of geological and spatial properties and also the complex fluids present in the reservoir. The data available to build a representative model are generally either sparse well data or poorly resolved seismic data. The resulting model is dependent on the structural complexity, the depositional model, the available data, and the objectives of the project. Building a usable reservoir model is always a compromise: we are trying to represent the reservoir not replicate it.

The advances in computer processing power and graphics over the last 20 years has meant that geoscientists can build representative models of a reservoir to capture the variability present at all appropriate scales from the microscopic to the field-scale. However, as reservoirs are complex, we need to be highly subjective about the scale at which we model and the level of detail we incorporate: a gas reservoir may well be a tank of sand, but faults may compartmentalize that tank into a number of separate accumulations.

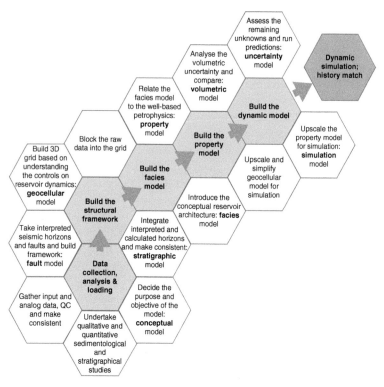

Assess the remaining unknowns and run predictions: **uncertainty** model

Analyse the volumetric uncertainty and compare: **volumetric** model

Relate the facies model to the well-based petrophysics: **property** model

Block the raw data into the grid

Build 3D grid based on understanding the controls on reservoir dynamics: **geocellular** model

Take interpreted seismic horizons and faults and build framework: **fault** model

Gather input and analog data, QC and make consistent

Dynamic simulation; history match

Build the dynamic model

Upscale the property model for simulation: **simulation** model

Build the property model

Upscale and simplify geocellular model for simulation

Build the facies model

Introduce the conceptual reservoir architecture: **facies** model

Build the structural framework

Integrate interpreted and calculated horizons and make consistent: **stratigraphic** model

Data collection, analysis & loading

Decide the purpose and objective of the model: **conceptual** model

Undertake qualitative and quantitative sedimentological and stratigraphical studies

Figure 4.2   Reservoir modeling workflow: the elements are presented as a traditional linear workflow showing the links and stages of the many different processes required to build an integrated reservoir model.

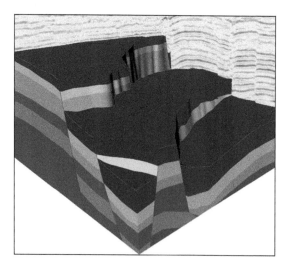

Figure 4.3   An example of a 3D reservoir model showing inclined faults and stratigraphic layering.

4.1.3   Available Resources

In my experience, the best field development teams need a geologist, seismic interpreter, and reservoir engineer working together in the same office around an integrated suite of software that allows data to be shared at all stages of the model build: this is called the parallel workflow model (Figure 4.4). Specialist services such as high-resolution stratigraphy, sedimentology, core analysis, petrophysics, quantitative geophysics, fluid analysis, and economics can be drafted in at the appropriate time either as in-house or external consultants depending on where the skill set lies. It is up to the core team and team manager to direct, control, and validate the work of the specialists. Sometimes the project may be limited by lack of data or budget, or the availability of the specialist service at a crucial point in the exercise, and a compromise will be made; however, it is important that any compromises are noted for future clarification when the resource (or data) becomes available.

Sometimes the team requires a specialist geo-modeler or simulation engineer if these skills are not available; in this case I believe that these specialists must have an equal voice in discussion and decision-making for the project. Having worked as a consultant in a number of internal and external projects I know that this is often a breakdown point. The

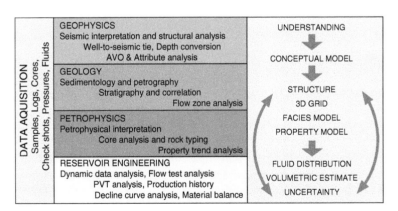

Figure 4.4   An example of the parallel workflow modeling adopted by many companies for an efficient, integrated approach to reservoir modeling. The separate disciplines work together using an integrated software solution to design the model, analyze the data, and understand the uncertainties. *Source:* Reproduced from Cannon. S.J.C. (2016) Petrophysics: a practical guide. John Wiley & Sons, Ltd. © 2016, John Wiley & Sons.

team manager may ask for the impossible result of managing the budget better or the seismic interpreter and reservoir engineer are at odds over the detail of a structural model, and the geo-modeler is in the middle of the debate with little opportunity to influence.

When it comes to the type of numerical simulator, the most common is a 'black oil' variety that is suitable for isothermal reservoirs containing immiscible oil, gas, and water. Numerically, a black oil simulator is defined as a finite difference, implicit pressure explicit saturation (IMPES) tool that finds the pressure distribution for a given time step first, and then calculates the saturation distribution for the same isothermal time step. Compositional simulators are used for reservoirs where the hydrocarbon phase compositions and properties vary significantly with pressure below the bubble point or dew point of the fluids. Both types of simulator use a corner point grid; we will look into grids in more detail in Section 4.2. There are other specialist numerical simulators, such as thermal, where the temperature varies significantly with depth, or dual media simulators where the reservoir comprises two interconnected pore networks such as a matrix and fracture combination.

It is important that the team have a plan in place before the project starts, and the plan must be based not only on the available data and resources but on the reason for building the model in the first place.

4.1.4 The Reason for the Model

Why are we building this model? What question are we hoping to solve by building the model? How long have we got to do the work and what is realistically achievable in the time available, with the tools and personnel allocated?

An integrated reservoir model should be built to answer a specific aspect of the subsurface that impacts on hydrocarbon distribution or fluid flow, which in turn has an impact on reservoir management. When designing a reservoir model, the ultimate purpose of the model must be defined: should it address the structural uncertainty of a prospect, the distribution and connectivity of reservoir units, or perhaps the definition of a number of infill well locations? Each of these challenges will require a different approach and functionality; however, the key will be flexibility of both the ideas and the solutions that are generated by the modeling team.

The type of model to be built should be a function of the question we are trying to answer. In gas fields a simple 'tank' model will often suffice, but if you are trying to capture horizontal and vertical heterogeneity a 3D model is required. However, simple 1D models can be used for upscaling studies, often needed when moving from the static geo-model to the simulation; cross-sectional models can be used to study displacement processes and radial models used to look at well-production behavior.

4.2 Designing the Modeling Grid

The model grid should be a key element of the whole modeling process (Figure 4.5); it is, however, a compromise between detail of content and scale, and computer run time. There are five key issues to take into consideration when constructing a grid for dynamic simulation:

1) *Geological issues*: The type and scale of any heterogeneities to be captured.
2) *Dynamic issues*: Specifically, the impact of permeability on vertical and horizontal flow with respect to different fluids.
3) *Numerical issues*: Due to the orientation of the grid and the subsequent numerical dispersion.
4) *Grid geometry*: Is the grid construction based on a Cartesian, corner point, or hybrid structure?
5) Total number of active cells in the model and their connectivity and location.

There are two basic geometrical designs for a model grid: the Cartesian grid system beloved of reservoir engineers because it is block centered and results in a fully orthogonal grid, and the corner point grid system that allows for a more accurate representation of the geological features of the field. In a corner point grid, the co-ordinates of each grid block are specified instead of the centers.

The corner point grid is defined as a combination of the UTM co-ordinates and true vertical depth (x, y, and z). The Cartesian co-ordinate system uses the number of the cells in each direction of the grid (i, j, k). This system has its own defined origin independent of the UTM system. The local directions of this system will vary with the local directions of the grid lines. The (i, j, k)-co-ordinate system should be

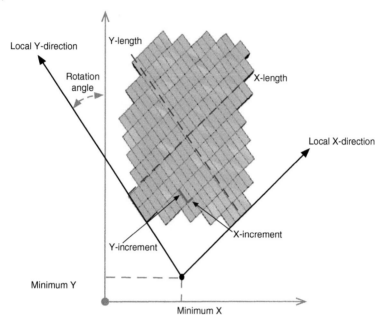

Figure 4.5 Grid orientation and axes nomenclature used in geocellular modeling. *Source:* Cannon. S.J.C. (2018) Reservoir Modelling: a practical guide. John Wiley & Sons, Ltd. © 2018, John Wiley & Sons. Reproduced with permission of Emerson-Roxar.

defined so the axes form a right-handed system; defining the origin in the north-western corner of the model can achieve this. The origin should be placed in the same corner in the geological and in the simulation model, to avoid confusion. The strict definition of the axis system is enforced to avoid having to redefine the grid and the properties later in the modeling process.

When a team is truly integrated it is possible for them to decide, on the basis of the objective of the model, all of the above. It is then possible to design the final simulation grid and thus mitigate many of the problems associated with upscaling properties, cell size, and cell layering. The geological model can always be more refined if required. Never forget that the model is based on representative data that is many times removed from either the field as a whole or the scale of the pores in which flow takes place.

4.2.1 Structural and Stratigraphic Modeling

The structural model is built from the depth-converted seismic horizons and fault data, generating a reservoir framework. This is combined with the internal reservoir layering that incorporates the stratigraphic component of the model. The structural model is often designed with a gross tectonic interpretation in place reflecting the interpreter's understanding of the regional structural history; in extensional basins normal faulting is expected, whereas in compressive settings reverse faulting and slumping might be predicted. Understanding the gross depositional environment of the interval drives the internal layering of the reservoir zones, leading to stratigraphic correlation and hierarchy. With the structural and stratigraphic models defined, the fine-scale geocellular model can be created.

The ever-improving quality of seismic acquisition processing methods means that seismic interpreters can delineate more and more detailed faulting over a field. It makes little sense to the reservoir modeler to capture all of these faults especially as the dynamic model will probably not be able to replicate this detail. Therefore, a hierarchy of fault importance should be established when building the geological model so that a realistic set of faults is represented; these faults should be seen to have an impact on the hydrodynamics of the field or the volumetrics. The following criteria could be applied in choosing which faults to model:

- Bounding faults that delimit the reservoir.
- Faults that divide the reservoir into separate compartments.
- Faults that intersect wells and are therefore considered hard data.
- Other seismically resolvable faults that might impact fluid flow.
- Other definable faults might be handled as transmissibility barriers.
- Fractures that are recognized in core or logs might be represented as stochastic permeability modifiers.

A detailed structural analysis of the field area should also be carried out at this stage to define the types of fault present and their relationships, especially if some gross conceptual structural interpretation is preferred.

The reservoir framework is completed by the addition of stratigraphic levels represented by seismically interpreted horizons/events and geologically significant surfaces identified in well data: where the levels are identified in both datasets then the mapped seismic horizons are

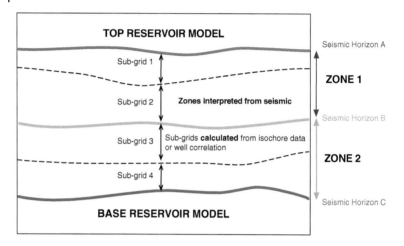

Figure 4.6 Horizon, zone, and sub-grid nomenclature used in geocellular modeling. *Source:* Cannon. S.J.C. (2018) Reservoir Modelling: a practical guide. John Wiley & Sons, Ltd. © 2018, John Wiley & Sons.

constrained by the well picks. All that is needed to build the reservoir framework is a top reservoir horizon and a base, ideally derived from seismic interpretation. Internal stratigraphic levels are usually calculated from correlatable horizons seen in the well data; these are often incorporated as isochore thickness maps (Figure 4.6). The internal zonation should reflect major changes in geology that have some influence on flow in the reservoir. The changes could be in the gross depositional environment, or in the type or style of heterogeneity or a change in the facies type or orientation. Getting the number of zones right will ultimately help build a robust 3D geocellular grid.

Selection of the number of horizons to use is a key component of the modeling process: in general, the fewer horizons the better. It is generally quicker to model at a coarse scale first to determine the level of detail required to achieve the objective. This step in the structural model requires the modeler to specify the type of horizon being modeled: whether is continuous across the model, truncated by another erosive horizon, or if it forms a base of a sequence that by definition builds upwards (Figure 4.7). These choices reflect the gross depositional setting of the reservoir sequence and the conceptual model. It is often necessary to test the relationships in the model to see if the correct sequence of events is realized.

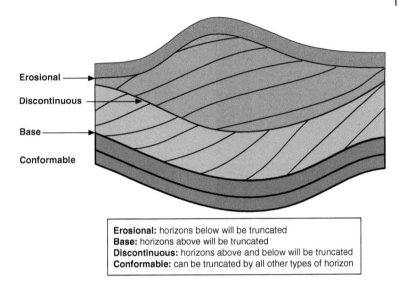

Erosional

Discontinuous

Base

Conformable

> **Erosional:** horizons below will be truncated
> **Base:** horizons above will be truncated
> **Discontinuous:** horizons above and below will be truncated
> **Conformable:** can be truncated by all other types of horizon

Figure 4.7 Classification and impact of different types of horizons used in reservoir modeling. *Source:* Cannon. S.J.C. (2018) Reservoir Modelling: a practical guide. John Wiley & Sons, Ltd. © 2018, John Wiley & Sons. Reproduced with permission of Schlumberger-NExT.

Associated with the choice of horizon type is the decision to use a lithostratigraphic approach to well correlation or to employ sequence stratigraphic principles (Figure 4.8). The decision may depend on the level of geological interpretation and complexity of reservoir architecture. Generally, a chronostratigraphic approach will integrate the seismic horizons more completely into the model, and therefore a sequence-based zonation will be more effective. The correlation of well-derived correlatable events will influence reservoir body connectivity; a petrophysical approach might only correlate sand bodies or flow zones in two dimensions and ignore potential changes of facies between wells. All of these decisions depend on the level of understanding of the reservoir framework and what is required to represent geology in the model.

The structural and stratigraphic models provide the skeleton of the reservoir framework and comprise the largely deterministic element. The geocellular model provides the fine-scale internal architecture of the reservoir that will ultimately be populated with facies and petrophysical properties. The definition of the internal architecture should not be done

North   LITHOSTRATIGRAPHY   South

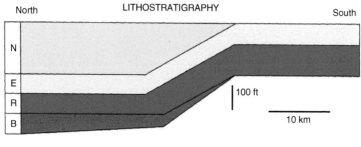

SEQUENCE STRATIGRAPHY

Figure 4.8 Comparison between lithostratigraphic correlation and sequence stratigraphy of the Brent Group, fluvio-deltaic system, North Sea. *Source:* Wehr and Brasher (1996). Reproduced from Cannon. S.J.C. (2016) Petrophysics: a practical guide. John Wiley & Sons, Ltd. © 2016, John Wiley & Sons.

without reference to the conceptual depositional model and, to a large extent, the dynamic properties of the reservoir.

The volume between the deterministic layers is filled with cells that are designed to reflect the patterns of deposition of the sediments and also the size of definable facies associations. Thus, packages of sediments that onlap, offlap, downlap, or are truncated can be modeled in a vertical sense. If the shape and size of particular facies bodies is to be modeled, then these parameters should be used to define the cell dimensions. It is at this stage that the sedimentologist can best influence the design of the model.

Varying the *xy* dimension of the cell in a particular direction can capture evidence of flow anisotropy. The reservoir engineer should use existing dynamic data to help align the overall model structure with the primary flow direction. If the concept of flow units is to be adopted, then some idea of the controls on their distribution should be included in the modeling grid.

4.2.2 Capturing Heterogeneity

Before building the final grid it is necessary to decide the scale of heterogeneity that impacts on fluid flow. This is a function of the geology and
the hydrocarbon fluids to be modeled dynamically. In general, this will
be dictated by:

- Does the available data resolve the heterogeneity at the scale intended?
- Will the grid block dimension properly capture the heterogeneity
  being modeled?
- Is there sufficient time to build a fine-scale model and is it required?
- Do you have access to appropriate modeling tools?
- What is the impact of the heterogeneity being modeled on flow?

In the end it is a matter of permeability, fluid fill, and production
mechanism; as permeability heterogeneity increases beyond one order
of magnitude, the need to build a 3D model becomes more necessary
(Table 4.1). Gas fields producing under natural depletion will often only
require a tank or material balance model unless the reservoir is highly
layered or the development requires horizontal or designer wells. An oil
field undergoing water injection should be modeled if the permeability
varies by two or more orders of magnitude (Ringrose and Bentley 2015).

Different heterogeneities can be effective at different scales and impact
reservoir connectivity, sweep efficiency, both laterally and vertically, and
the rock fluid interaction (Figure 4.9) (Weber 1986). At the largest scale,
sealing and semi-sealing faults can have a great impact on both reservoir
connectivity and sweep efficiency. Facies boundaries can act as barriers
to flow when the permeability contrast is significant; the change between

Table 4.1 The requirement for building a 3D reservoir model increases
with permeability heterogeneity, fluid type, and production mechanism.

Production mechanism	No aquifer support	Aquifer support	Water flood	Gas/steam flood
3 OM K	(Model)	Model	Model	Model
2 OM K	No model	(Model)	Model	Model
1 OM K	No model	No model	(Model)	Model
Fluid	Dry gas	Condensate	Light oil	Heavy oil

Heterogeneity		
Giga (>1000 ft)	Sealing fault	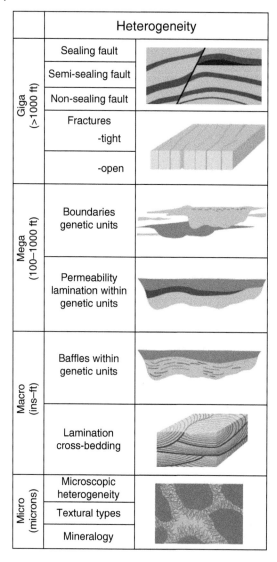
	Semi-sealing fault	
	Non-sealing fault	
	Fractures -tight	
	-open	
Mega (100–1000 ft)	Boundaries genetic units	
	Permeability lamination within genetic units	
Macro (ins–ft)	Baffles within genetic units	
	Lamination cross-bedding	
Micro (microns)	Microscopic heterogeneity	
	Textural types	
	Mineralogy	

Figure 4.9 Types of heterogeneity at different scales from the microscopic to the basin. *Source:* Weber, K. J. (1986) How heterogeneity affects oil recovery. In: Reservoir Characterization. Lake, L. W. and Carroll, H. B. J. (eds.) Academy Press: 487–544 ; Reproduced from Cannon. S.J.C. (2016) Petrophysics: a practical guide. John Wiley & Sons, Ltd. © 2016, John Wiley & Sons.

a permeable sand-filled channel and the relatively impermeable flood-plain deposits. Internal sedimentological layering and lamination may act as baffles to flow, reducing sweep efficiency during both aquifer ingress and water flooding. The rock fluid interaction is a function of pore geometry and mineralogy affecting wettability and capillary pressures in the reservoir: it is at this microscopic scale that flow in the reservoir actually occurs. A few research-driven operators are prepared to model the genetic elements of the reservoir at a sufficiently small scale to try and capture the impact of laminations and pore-scale heterogeneity.

The large-scale architecture of reservoir and non-reservoir units is a major control on hydrocarbons initially in-place, drainage, and sweep efficiency. This has an impact on the volume of hydrocarbons that can be recovered. Drainage and sweep efficiency are governed by the connectivity of the reservoir units. This is primarily governed by the net:gross (NTG). The identification of genetic reservoir units (such as channels, bars, etc.) is a key step in modeling the large-scale reservoir architecture (Weber and Geuns 1990). These units form the basis of object-based reservoir modeling techniques (Figure 4.10). The impact of large-scale reservoir architecture on flow depends not only upon the nature of the architecture, but also upon the fluid properties and flow regime. The impact of heterogeneity on flow is accentuated if the mobility ratio is unfavorable.

In summary, keep the framework as simple as possible to ensure the most robust grid for simulation; avoid hand building the grid, it is easier when it comes to updating the model; use model segments and zones to construct different reservoir volumes from the start as it is difficult to change the framework later; capture the large-scale heterogeneity that affects fluid flow and think about upscaling and down gridding strategies. A robust geological grid with assigned properties is the most complete representation of the geological model and may be used for volume calculations and, in some cases, for well planning. This model will usually be upscaled for dynamic simulation.

The geological grid may be faulted (corner point) or non-faulted (XY regular), but since faults are so important for many reservoirs, the general recommendation is a faulted grid. Hence, the rest of this section will focus on faulted grids. If the upscaled geological grid with properties is to be input for reservoir simulation, it is very important that the reservoir engineer is involved in grid design.

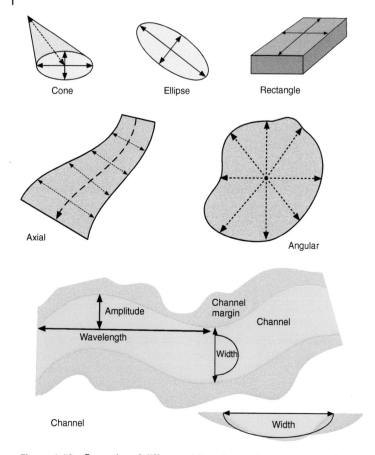

Cone   Ellipse   Rectangle

Axial

Angular

Channel

Amplitude

Channel margin

Channel

Wavelength

Width

Channel   Width

Figure 4.10   Examples of different object shapes that can be modeled to represent genetic bodies or facies. *Source:* Cannon. S.J.C. (2018) Reservoir Modelling: a practical guide. John Wiley & Sons, Ltd. © 2018, John Wiley & Sons.

The main product is a grid that represents reservoir geometry and is the basis for modeling of geological properties. The form of this product depends critically on the planned use of the gridded model. There are two endmembers when it comes to gridding, depending on what the model should be used for:

- If the main purpose of the model is well planning or volume calculations, an accurate representation of faults is essential. The grid should then be constructed with sloped faults and irregular grid cells shaped after the exact position of the faults.

- If the main purpose of the model is construction of a reservoir simulation model, the geological grid and the simulation grid should be constructed together to minimize sampling errors in the upscaling from fine scales to the simulation scale.

It is possible to combine the purposes, being aware that the geological model created may be suboptimal for both volumetrics and upscaling to simulation level. Such a pragmatic approach can be defended in cases where accuracy around faults is of minor importance. The grid should include at least all the faults that will be included in the simulation model. Faults in the simulation grid should use the same naming convention as in the fault model.

Ideally, the axis of the geological grid should be aligned with main direction of the controlling influences on flow in the reservoir. This will usually be the major faults or the primary depositional facies. Sometimes grids may be aligned with the direction of horizontal wells. The axis of the grid should define a right-hand co-ordinate system, with the origin defined in the 'north-western corner' of the grid. As with grid construction in general, orientation of the grid is a compromise between several different factors of both input data and results.

4.2.3 Orientation from Seismic Lines

Seismic data will be collected along seismic lines with a global direction relative to the field. If data from the seismic cube are to be used to condition the geological model, it is preferential to orient the geological grid in the same directions as the survey to minimize sampling errors. The cell length in the direction normal to the seismic lines should then also equal the distance between the seismic lines. Ideally, the grid should then be completely regular.

4.2.4 Orientation from Major Faults

If an exact representation of faults is attempted in the grid, cells will always be somewhat distorted in the vicinity of faults. Stochastic modeling assumes that all geological grid cells are close to equal size, and an error is introduced if grid cells are deformed too much. This error is reduced if the grid can be constructed to follow major fault directions, so that the number of grid cells that are deformed to fit the fault pattern is minimized (Figure 4.11).

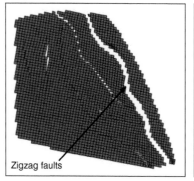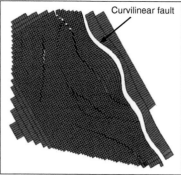

Figure 4.11   Grid resolution aligned to the major structural features creates a more robust and realistic faulted grid. *Source:* Cannon. S.J.C. (2018) Reservoir Modelling: a practical guide. John Wiley & Sons, Ltd. © 2018, John Wiley & Sons. Reproduced with permission of Emerson-Roxar.

4.2.5   Orientation from Geological Features

The geological data will generally reflect some degree of anisotropy, for instance, in properties such as permeability, variogram range, and so on. This anisotropy usually reflects the major trends in the depositional system, for instance, parallel and normal to the main channel directions in a fluvial system. Geological anisotropy must be modeled based on the local axis in the (i, j, k) system in the geological grid. The direction of the geological grid should then be selected to coincide with the characteristic geological directions. A common problem is, however, that geological orientation changes from zone to zone. And it is almost impossible to have grids with varying directions for each zone. The recommendation is then to follow the geological features in the most important reservoir zone(s) (Figure 4.12).

4.2.6   Orientation from Simulation Grid

Sampling errors typically occur when properties are upscaled from the geological grid to the simulation grid. This error is minimized if grid cells in the geological grid and the simulation grid match each other. Construction of a simulation grid usually involves flow considerations (e.g. following wells), which is not an issue for the geological grid. To create a correspondence between the two grids, these considerations should also be taken into account for the geological grid (Figure 4.13).

Grid aligned with fault orientation

Grid aligned with channel direction

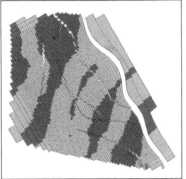 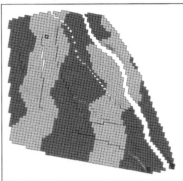

Figure 4.12 When a grid is aligned to the primary flow direction, a better grid for dynamic simulation is created. *Source:* Cannon. S.J.C. (2018) Reservoir Modelling: a practical guide. John Wiley & Sons, Ltd. © 2018, John Wiley & Sons. Reproduced with permission of Emerson-Roxar.

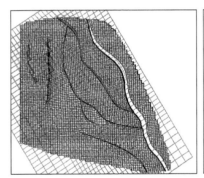 

Figure 4.13 The SmartModel concept promotes building the geocellular and simulation grids with the same orientation and complementary dimensions so that upscaling and down gridding methodologies can be improved. *Source:* Cannon. S.J.C. (2018) Reservoir Modelling: a practical guide. John Wiley & Sons, Ltd. © 2018, John Wiley & Sons. Reproduced with permission of Emerson-Roxar.

The various issues influencing orientation discussed in Sections 4.2.2–4.2.5 are often in conflict with each other. If a ranking of these conditions is needed, orientation from geological features and simulation grid are most important. In addition, there is a more practical consideration, and that is to minimize the number of cells. Therefore, in order to optimize computation speed and visualization speed, it is advised to select a direction that makes the geological grid as small as possible in terms of numbers of cells.

4.2.7 Cell Sizes and Total Number of Cells

Ideally, grid size should be selected small enough to capture the smallest geological feature required in the model. The Nyquist sampling theorem (used in geophysics) suggests that the grid cell should not be larger than half the length of the smallest feature, to avoid aliasing (erroneous sampling). This typically leads to a vertical resolution of 0.5–2 m grid cells. Aliasing is a phenomenon that occurs when an analog object is sampled to a digital representation. If the sampling is performed with too low resolution, the objects are not rendered correctly and large distortion may occur in the digital representation.

Unfortunately, the areal grid size typically must be a compromise according to the total machine memory available for the model. The total number of cells depends on machine capacity and the patience of the user. This will change from year to year as hardware is improved. By using a prototyping procedure to test the model construction, a much smaller grid can be initialized; this can be done by selecting a sector of the reservoir. At an initial stage, test the modeling workflow on a grid with less than 500 000 cells.

The geometry of a simulation grid can vary depending on the model purpose (Figure 4.14):

- 1D models are used for sensitivity analyses to test specific reservoir parameters or to look at upscaling petrophysical properties.
- 2D cross-sectional models are used to study vertical displacement issues, such as the impact of flank water injection or crestal gas injection.
- Areal 2D models are built when lateral flow patterns dominate reservoir performance; typically, they are used for water flood pattern studies, where vertical heterogeneity is not an issue.
- Radial models are restricted to the area around a well bore and are used to evaluate the well performance when subjected to large vertical pressure gradients; commonly used to study gas and water coning or cusping in horizontal wells.
- 3D dynamic models are most commonly used for field performance and ultimate recovery predictions.

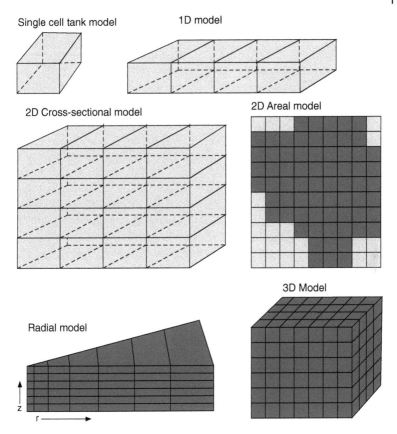

Figure 4.14 Different types of simulation grids for use in dynamic flow investigation.

4.3 Facies Modeling

A facies model captures the medium-scale reservoir variability based on the sedimentological analysis of the core and wireline data, combined with a conceptual model of the reservoir depositional environment. The main reason to build a facies model is to condition the subsequent property model; each facies should have porosity and permeability distribution that is different from the other facies. This could be as simple as good sands, moderate sands, and poor sands. If the reservoir quality can be attributed to specific geological bodies or environments, then representative heterogeneity can be introduced into the model. Depending on the distribution of facies the modeler has choice of using *pixel-based* or *object-based* modeling methods: where the facies form a mosaic pattern, such as floodplain or carbonate shelf, pixel-based, indicator simulation may be used, whereas channel or shoal bodies might be characterized using object-based modeling. Either way, geology will be built into the model and reservoir connectivity might be captured better, as well as populating non-cored sections of a well.

Facies modeling is essentially the filling of the fine-scale geological grid with petrophysical properties. Modeling discrete facies is a way to improve the petrophysical model, and is an essential task where reservoir property distribution is a function of the sedimentary depositional system. Selection of the basic modeling parameters is a combination of detailed sedimentology and regional depositional models. The modeling procedures are very much dependent on the available software routines and on the complexity of the modeled reservoir volume. The reservoir properties can be modeled using deterministic, stochastic–deterministic, or stochastic methods with all well observations being honored in the process. The modeling process at this stage requires strict quality control both visually and quantitatively.

Using the conceptual model plus well data the sedimentologist can provide a facies breakdown to be modeled. Turning this detail into a usable scheme for 3D facies modeling requires two key steps: first in analysis and second in application:

1) Qualitative and quantitative analysis of trend information from logs and well data. This should be used to give a deterministic understanding (low frequency) of parameter variation such as facies proportions and flow properties; any observed trends should be explained sedimentologically. This process can be used as part of the well data quality control. Any vertical and linear trends in facies or property data should be recognized and explained.
2) Perform blocking of well data; the first step in upscaling from raw log data: there are usually a number of methods with which to block the wells. Conditioning on dynamic data should be delayed to a history match of the simulation model, but well test data could be used for quality control. Perform quality checks before further modeling including a comparison of raw and blocked data.

4.3.1 Defining the Facies Scheme

The first step in the workflow is to decide in what depositional environ-ment the reservoir was deposited: continental, transitional, or marine. In each of these gross environments we can predict the type of geological bodies likely to be present, their distribution, and their associated facies. This should all have been considered in developing the conceptual model. Detailed core descriptions and calibrated wireline log interpreta-tions should identify the types of facies present in the well data and their vertical and lateral relationships: this amounts to building the deposi-tional model on paper, and if you can draw your model, then you will be able to build a 3D geocellular representation. Where well data are limited it may be sufficient to build a simple sand/shale model based on a gamma-ray cut-off, but some idea of the proportion and distribution of each facies should be attempted if only to test the uncertainty in NTG.

4.3.2 Upscaling of Log Data (Blocking Wells)

Before the well data can be used for modeling, it must be scaled-up to the vertical resolution of the 3D grid; this applies to both facies and property data. Blocking of wells is the term used for the process of upscaling from log scale to the geological grid scale. The facies property is a discrete ran-dom variable: it is a finite property ranging from $0. . .n$, where n is a whole number, hopefully not greater than 5! Raw log data are normally sampled every 6 ins (15 cm), grid cells are usually larger in the vertical direction. A note of caution: remember that you are performing an upscaling routine; this means that data will be averaged out and impor-tant data can be lost, such as thin high permeability streaks or thinly bedded shales, so visually inspect the results.

Preparation of the log data before the upscaling must reflect the selected method for handling non-reservoir facies types. If, for instance, it is intended to model discrete calcite nodules or dense non-reservoir zones independently, then all such observations must be excluded from petrophysical analysis characterizing the other facies types (or zones).

Blocking of wells should be performed as follows:

- Facies data should be pre-processed to remove shoulder effects.
- A zone log must be defined, that is, a log containing the zone index of the intersecting zone along the well path.

- When blocking the zone log it may be necessary to 'shift and scale' logs to match the sub-grids; QC the result for unreasonable depth shifts.
- Cell layer averaging is recommended for all continuous variables (porosity, permeability, etc.).
- For discrete logs such as a facies log use the 'most of' routine.
- Make sure that facies that are less than a single cell thick are captured as a discrete package; this is particularly true for high permeability streaks and carbonate nodules if they make up a significant (5–10%) proportion of the well sequence.

4.3.3 Simplified Facies Description

In some cases, a detailed facies definition may not be required for the modeling purposes: perhaps a simpler facies description can be used than the one developed as part of the depositional analysis. The following list describes cases where simplifications are reasonable:

- If the zone is a non-reservoir zone, a zonal average model may represent a sufficient simplification. The modeling becomes similar to conventional layer average modeling. Be aware that small errors in grid-to-well connection may contaminate such a model with wrong observations.
- If the zone mainly represents good upward coarsening or upward fining units with no apparent or expected discontinuities, such trends may be modeled directly in a petrophysical model, that is, creating an effective model by simply bypassing the facies model.

With this approach, trend analysis of the petrophysical distributions becomes more important as trends are no longer captured in the facies proportions. The petrophysical analysis becomes more demanding. For geological environments where facies transitions are characterized by a strong contrast in some log response, the estimation of the vertical variogram ranges may be interpreted from such logs. The horizontal variogram normally has to be chosen in a more pragmatic way based on geological knowledge.

4.3.4 Facies Modeling Methods

Turning a set of logs, 2D maps, and a conceptual model into a visual representation that can be viewed in 3D is for many geologists the pinnacle of their technical career; taking all that they learned in school and university and doing real geology again. But to achieve a reasonable result that makes both geological and geostatistical sense is a huge challenge and one that should not be undertaken lightly. Choosing the right method to distribute facies is an important element in the process, and one that often only comes with experience. Section 4.4.1 briefly summarizes the different techniques available straight out of the 'toolbox'; they fall into two categories, pixel- and object-based.

4.3.5 Pixel-Based Methods: Indicator and Gaussian Simulation

Pixel-based models are built using correlation methods based on the variogram, a measure of spatial variation of a property in three orientations: vertical and maximum and minimum horizontal directions. Experimental variograms can be fitted as spherical, normal (Gaussian), and exponential trends to the input data to produce different outcomes of the facies or property model. These 'models' can then be used to distribute a property correlated to trend or a seismic attribute. Indicator simulation is used for discrete properties and Gaussian simulation for continuous properties.

4.3.6 Indicator Simulation

In most geological settings, rock properties can be grouped into populations that are genetically related. These genetic (depositional) units have geometries and are, themselves, spatially correlated. Indicator simulation methodologies (Figure 4.15) utilize simple kriging to estimate the probability of a facies transition utilizing indicator variograms. These methods are most applicable when the density of well information is greater than the average size of facies objects, and often requires hundreds of regularly spaced wells. Where a robust relationship exists between a seismic parameter (acoustic impedance) and facies (sand or shale), indicator simulation can return a realistic looking result, but the need to calibrate the model is essential.

In sequential indicator simulation (SIS) an indicator variable (discrete) is selected from a preconditioned distribution of values to represent a proportion of a given facies. Indicator simulation is easy to use, but the variogram can be difficult to interpret (model). It is a flexible and data-driven modeling tool; however, the results are inexact; it does not reproduce the variogram or histogram exactly and does not give a very geological looking outcome.

Figure 4.15   An example of an indicator simulation model of an alluvial floodplain comprising shales (green), overbank deposits (brown), and channel sands (yellow). Note that the channel sands are not continuous ribbons across the model. *Source:* Cannon. S.J.C. (2018) Reservoir Modelling: a practical guide. John Wiley & Sons, Ltd. © 2018, John Wiley & Sons.

Sequential simulation techniques explicitly calculate the conditional distribution at each point; the user then samples from this distribution.

Sequential Gaussian simulation is used to distribute continuous variables and will be discussed in more detail in Chapter 5; a special case of Gaussian simulation used for facies modeling is discussed in Section 4.4.3.

4.3.7   Truncated Gaussian Simulation

Pixel-based methods for modeling transitional facies are based on Gaussian fields, and implement a truncated Gaussian simulation algorithm (TGSim, Figure 4.16). This method applies a Gaussian field, that is, a continuous field, and thereafter truncates it into discrete facies. This implies that the gridding technique (direction, vertical grid setup) is important for the result.

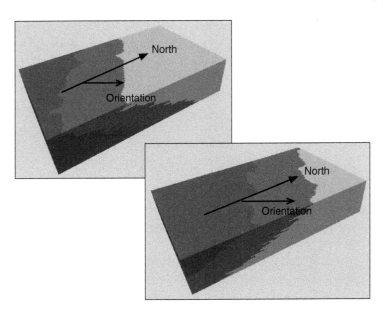

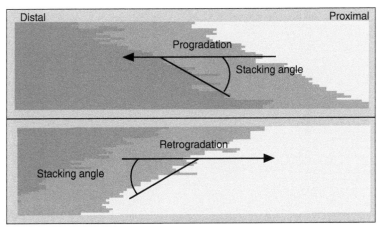

Figure 4.16 An example of truncated Gaussian simulation using trends to model progradation and retrogradation of shallow marine sandstones. *Source:* Cannon. S.J.C. (2018) Reservoir Modelling: a practical guide. John Wiley & Sons, Ltd. © 2018, John Wiley & Sons.

TGSim requires a strict sequential setup of facies: if three facies are present (A, B, C), then facies A must be next to facies B, and facies B must be next to facies C. A conditioning point (a well) that has facies A next to facies C will not be accepted by the algorithm. It is important to establish a good input or 'prior' trend in TGSim that can then be tested against the results. TGSim is best used to model large-scale features such as progradational or retrogradational packages. In carbonate reefs TGSim can be used to model the transition from reef core to margin.

4.3.8 Object-Based Methods

Object- or Boolean modeling is an alternative method commonly found in the software products. There are generally two object-based methods for modeling: one is focused on fluvial reservoirs (Figure 4.17), while the other is a more general method that can model a wide range of flow-unit shapes (Figure 4.10). One key attribute of object modeling is that it

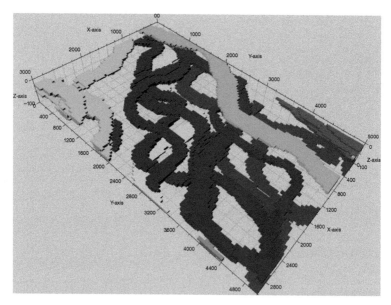

Figure 4.17 An example of fluvial channel objects to model connectivity: most channels are isolated, only those colored as pink are well connected in the model. *Source:* Cannon. S.J.C. (2018) Reservoir Modelling: a practical guide. John Wiley & Sons, Ltd. © 2018, John Wiley & Sons.

implicitly models connectivity of the bodies leading to an improved recovery factor in the dynamic model; indicator simulation cannot do this normally resulting in lower recovery.

Object-based modeling allows the user to build realistic representations of large-scale geological units such as channels, lobes, dunes, sand bars, and reefs. Understanding the shape and size of these bodies requires analog information gathered from outcrop or published literature; many companies have internal databases containing this information (Figure 4.18). When modeling specific objects, it is even more important to get the grid cell dimensions correct; there is no point in trying to model channels 50–100 m wide if the lateral cell size is 200 m. In object-based methods, one facies is regarded as background facies. These facies will not have any particular shape. Usually the facies that has the greatest proportion from well data is set as background; however, other 3D volumes/models can be used to form a composite model. The algorithm

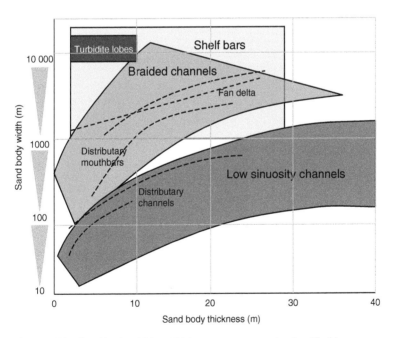

Figure 4.18 Sand body width to thickness measurements classified by depositional environments: data compiled from numerous outcrop locations. *Source:* Cannon. S.J.C. (2018) Reservoir Modelling: a practical guide. John Wiley & Sons, Ltd. © 2018, John Wiley & Sons.

works best if the fraction of background facies is 'large' (say, more than 30%), otherwise, it may be difficult to achieve convergence.

4.3.9 Conditioning to a Seismic Parameter

Conditioning facies to seismic can be done within either modeling algorithm, including multipoint statistical methods (MPS). However, using seismic to condition can be challenging and is very dependent on establishing a robust relationship between a property to be modeled and the seismic parameter. The workflow includes:

- Rescaling the seismic cube into the structural model.
- Blocking wells.
- Estimating the 'seismic' factor (so-called G-function) that makes a correlation between the seismic property (e.g. impedance) and facies (e.g. channel belts).

The relevant software manuals will give the necessary means to establishing G-functions.

4.3.10 Conditioning to Dynamic Data

At present, no validated method exists for upscaling log data conditioned on well test data. However, there are a number of qualitative and semi-quantitative tools that can be used to help in conditioning:

- PLT (production log data); simple spinner data which indicate intervals of major dynamic contribution to flow.
- RFT/FMT pressure data can indicate differential depletion or pressure communication between reservoir bodies.
- Tracer data and interference tests can indicate inter-well communication.

Dynamic data are currently only used as a quality control method; further matching between the static grid data and the dynamic well test data should be performed as part of a history match with the simulation model.

4.3.11 Flow Zones

A flow zone is a mappable unit of a reservoir having consistent geological and petrophysical parameters affecting fluid flow. The properties of a flow zone should be predictably different from other reservoir rock volumes. In essence, the way in which a reservoir is zoned for modeling

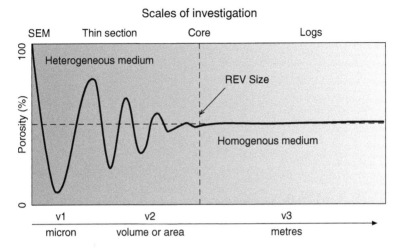

Scales of investigation

Figure 4.19 The representative elementary volume (REV) concept and the scales of investigation and measurement in heterogeneous and homogeneous media. *Source:* Bear, J. (1972) Dynamics of Fluids in Porous Media. New York: American Elsevier; Reproduced from Cannon. S.J.C. (2018) Reservoir Modelling: a practical guide. John Wiley & Sons, Ltd. © 2018, John Wiley and Sons.

could lead to defined flow zones; however, this is likely to produce simplified layer-cake models. A hydraulic flow zone is a dynamic version of the geological flow zone that introduces the concept of the representative elementary volume (REV) (Bear 1972) that recognizes the scale of flow from the capillary to the sequence (zone) (Figure 4.19).

To a geologist a flow zone is a definable facies object such as a fluvial channel or shallow marine sandbar; to a petrophysicist it is a correlatable zone with similar porosity, permeability, and NTG ratio; to a reservoir engineer it is a layer in the reservoir that has a consistent dynamic response in the simulator; to a reservoir it is all of these things. Recognizing a flow zone is a function of the petrophysical properties as much as the geology and is tied up in a discussion of rock types that is best left to Chapter 5.

4.4 Property Modeling

I prefer to use the term 'property modeling' rather than 'petrophysical modeling' because that has a different connotation for petrophysicists. Property modeling is about capturing the fine-scale distribution of porosity, permeability, and water or hydrocarbon saturation in the geocellular model. In fact, only porosity should be stochastically modeled as permeability is usually a function of porosity or rock type, and water saturation should be distributed through a height above free water level relationship. Where a robust facies model exists then the reservoir properties should be directly related to reservoir architecture.

The primary purpose of a 3D property model is to improve the understanding of hydrocarbon distribution for volumetric analysis. A facies-constrained property model tries to capture the heterogeneity in the reservoir in such a way that dynamics of fluid flow can be modeled more realistically. Generally, 2D maps of reservoir properties are smooth interpolations of the inter-well value, but we know that the subsurface geology is not smooth. The introduction of 3D property modeling means that the vertical variation in properties is preserved and used to populate the inter-well volume using simple geostatistical methods such as kriging. Other trends can also be superimposed on the model, such as simple porosity depth relationships or saturation-height functions.

In property modeling we are basically distributing the results of a petrophysical interpretation based on indirect measurement of physical responses of rock and fluid.

What are we trying to measure?

- *Porosity*: total or effective, primary, secondary
- *Saturation*: hydrocarbon, water, irreducible
- *Permeability*: absolute, effective, relative, movable fluids
- *Lithology*: sand, shale, limestone, dolomite

What measurements do we make?

- *Density*: rock and fluids
- *Acoustic*: rock and fluids
- *Resistivity*: rock and fluids
- *Nuclear*: rock and fluids

None of the measurements are direct: they are all indirect values based on interpretation of a physical response to a static property or some source of stimulation. The measurements must be calibrated by core analysis data, which is a direct, if imperfect, measurement, before they can be used for estimation of the desired property we hope to model.

Two other topics that need to be discussed are whether we are working in a *total* or *effective* property domain, and the use of NTG ratio in defining what contributes to volume or flow. The reservoir model should ultimately represent *effective* properties if you want accurate volumetric results, and a robust facies model should negate the need to apply cut-offs in defining what is *net* reservoir. The total property model (Ringrose 2008) is a potential solution to the problem if there are sufficient data.

4.4.1 Total vs. Effective Porosity

Total porosity is the total void space in a volume of rock whether it contributes to fluid flow or not: effective porosity excludes isolated pores and pore volume associated with water adsorbed on clays or other grains; in other words immobile water. Effective porosity is generally less than total porosity, but in clay-free, clean sands the two properties are likely to be similar. Petrophysicists calculate total porosity from wireline data using a number of different methods and calibrate the results with core-derived porosity measurements. After cleaning and drying the core samples the results returned are somewhere between total and effective porosity: the harsher the cleaning and drying process, the nearer to total porosity. A correction for overburden pressure can be applied to replicate the conditions in the reservoir.

What we are trying to model is the effective porosity in the reservoir at reservoir conditions to correctly estimate the volume of movable hydrocarbons in the field and a recovery factor. If we use total porosity, we will overestimate the in-place hydrocarbon volume. Discounting the clay-bound water using a volume of shale cut-off derives effective porosity. Ask the petrophysicist to provide this input for each well in the dataset, rather than having to calculate it yourself: sometimes it is easier being the petrophysicist and modeler combined!

Total property modeling (TPM) is an approach where all the rock properties are explicitly modeled and cut-offs only applied if required after modeling (Ringrose 2008). In this case, neither cut-offs nor NTG are part of the process (Figure 4.20). Because good-quality rocks are modeled alongside poor-quality rocks it is possible to test the impact of different cut-offs on the in-place hydrocarbon volume. The main advantage of TPM is that the link to the dynamic model is more easily established

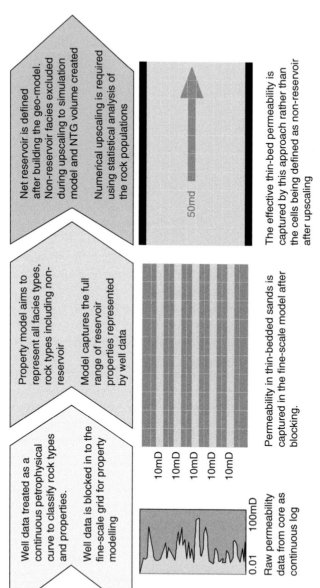

Well data treated as a continuous petrophysical curve to classify rock types and properties.

Well data is blocked in to the fine-scale grid for property modelling

Property model aims to represent all facies types, rock types including non-reservoir

Model captures the full range of reservoir properties represented by well data

Net reservoir is defined after building the geo-model. Non-reservoir facies excluded during upscaling to simulation model and NTG volume created

Numerical upscaling is required using statistical analysis of the rock populations

0.01   100mD

Raw permeability data from core as continuous log

10mD
10mD
10mD
10mD
10mD

Permeability in thin-bedded sands is captured in the fine-scale model after blocking.

50md

The effective thin-bed permeability is captured by this approach rather than the cells being defined as non-reservoir after upscaling

Figure 4.20   Total property modeling (TPM) avoids the need to apply NTG until upscaling properties for simulation. *Source:* Ringrose, P.S. (2008) Total property modelling: dispelling the net-to-gross myth. SPE Reservoir Evaluation and Engineering 11, 866-73 ; Reproduced from Cannon. S.J.C. (2018) Reservoir Modelling: a practical guide.John Wiley & Sons, Ltd. © 2018,John Wiley and Sons.

because the sequence of upscaling from well to blocked data to geo-model is traceable and in the simulator all cells that do not meet the cut-off criteria can be set as non-reservoir.

4.4.2 Blocking or Upscaling Well Data

A 1 m-thick cell at the well will comprise one single facies but will need to capture around six (log sample interval 15 cm) porosity values that might all be from the same rock type or from more than one distribution. How we average this will depend on the property being upscaled; any number of methods are available in the toolbox. The use of 'most of' or the 'histogram' method that worked for a discrete property like facies will not work for continuous variables.

There is obviously a link between the scale and the type of heterogeneity. If cells of $50 \times 50 \times 2$ m successfully models geology, it is an underlying assumption that we are modeling on a scale represented by this volume. The data needed for such modeling should, in theory, be scaled to be representative for the geo-modeling scale. A simplified solution used in this context is to use some kind of an average of the samples.

Porosity is a relatively simple property to upscale; it is static, dimensionless, and volume related, so a simple arithmetic average will work. The results may be averaged biased to the facies by a weighting approach; this has the effect of smoothing the outcome for all the cells in the well but gives a more representative range for a given facies. Non-net values should be 'undefined' and not zero if NTG is to be modeled: shale porosity should be included in the continuous porosity if the facies is to be used to define net and non-net rock. In this way a pseudo-NTG is generated for each cell that reflects a lower average porosity honoring the non-net volume. It may be appropriate to block well data on a zonal basis to allow you to change the method in each zone.

A log-derived water saturation property in a well is another continuous property that is a function of the pore volume. It is an additive property that can be upscaled using a summation approach like porosity; it should not be biased to the facies log. The only reason to upscale water saturation is to compare the log-derived results with the subsequent saturation-height calculation.

Before attempting to upscale permeability, find out how the log was generated: commonly, it will be a log-linear relationship with porosity derived from core data. However, unless you have discussed a multi-facies relationship, the results may not be appropriate for use with a facies model. One observation commonly made is that permeability prediction using a simple relationship almost always underestimates the high permeability values and overestimates the lower end. There are a number of methods for blocking permeability and the default is usually a geometric average, but depending on the heterogeneity of the data an arithmetic or harmonic mean may be equally valid: try the different methods available and compare the results to ensure that the extremes are captured. Permeability is a dynamic property, so involve the reservoir engineer at this stage.

It is important to QC the results of the blocking process by comparing the petrophysicists' input data and upscaled output data. This is done by comparing sums, averages, and standard deviation of the two sets of data. This should be done by facies, zone, and field area if known trends in the data exist.

4.4.3 Statistical Property Modeling Methods

Geostatistical property modeling allows more control on the spatial statistics of the model and also allows for smaller-scale heterogeneity to be captured. Here, an understanding of the uncertainty is required or where strong trends are present in the data. This approach gives the user the possibility of generating multiple outcomes in a similar way to stochastic facies modeling. In all cases the input data must be normally distributed, and any trends removed; this is usually performed in a data analysis module that is part of the property model software.

Most geostatistical modeling methods are based around the properties of a variogram, which estimates the spatial distribution of a property as it varies in 3D space. At small separation distances (h), the value of the variogram function exhibits its lowest property value (z). If this value is zero, it means that the measurements are identical when taken at the same location. The semi-variogram value at zero separation distance is referred to as the *nugget*. As the separation distance increases, the difference between samples increases at a rate determined by the function itself. At some distance, called the *range*, further increases in the separation distance have no effect on the difference between measurements. The range is the distance beyond which measurements

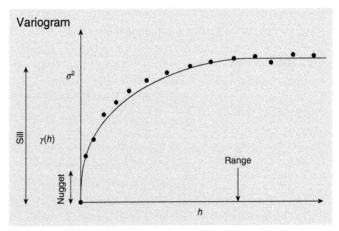

(a) Schematic diagram of a variogram with description of the main features - sill, nugget and range

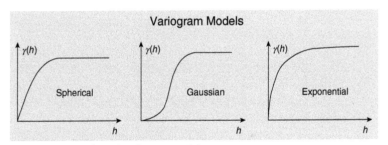

(b) Different experimetal variogram models

Figure 4.21 (a) Schematic diagram of an ideal variogram with an explanation of the main features: sill, nugget, and range. (b) Different experimental variogram models used in geostatistics. *Source:* Cannon. S.J.C. (2018) Reservoir Modelling: a practical guide. John Wiley & Sons, Ltd. © 2018, John Wiley and Sons.

are spatially uncorrelated. The z value of the variogram at h > range is referred to as the *sill*. This can be shown to be equivalent to the variance of the measurements themselves (Figure 4.21).

There are two main types of geostatistical methods:

> *Kriging* is an estimation technique that uses interpolation based on a weighted average of the existing data. A variogram is used to calculate those weights; kriging has the ability to incorporate

anisotropy as part of the weighting process. Another advantage of the kriging method is that it automatically declusters the data. Kriging creates smooth transitions from one point to the next; it does not replicate the variability that may exist at scales smaller than the space between the data points. Kriging estimation is always the expected value at the point; the variance at a point is never utilized.

Simulation is an alternative method that will avoid the smoothing tendency in the kriging methods. Most simulation algorithms utilize a randomly sampled value from the conditional distribution at a given point to estimate the property. This constrained randomness models the variance at all scales and therefore produces spatial distributions that can vary from realization to realization and yet always honor the data and the variogram. An experimental variogram still needs to be defined for any of the simulation methods.

Simulation does a much better job, visually, of representing the actual data than the kriged estimates; it honors the variogram, indicating that the information regarding spatial variance has been captured. As a result of honoring the spatial variance model, the distribution of values in the realization also matches the distribution of the input data; kriging does not honor that distribution.

Some commonly used methods used for property modeling based on simulation are:

- Sequential Gaussian simulation
- Sequential Gaussian simulation with external drift
- Sequential co-located co-simulation
- Gaussian random function simulation

Simulation and kriging conserve the spatial statistics in three directions (most importantly the vertical heterogeneity in permeability) as seen in the sample values, whereas in interpolation this variation is not taken into account – the variogram for interpolation would look like a straight line rather than a curve. Simulation captures the extremes of permeability and directional variability, whereas kriging only captures the directional variability: interpolation does neither. Ultimately, it is the degree of variability (heterogeneity) in permeability that controls fluid flow through a reservoir. Breakthrough time for a simulated model will be far more accurate than from an interpolated model.

The final geocellular model should always have the right balance between deterministic data, aspects of the model that are known by the modeler, and probability, those unknown knows that are specified by the modeling software. Whether a well has sampled all possible facies or how they are distributed is more often a function of the probability specified by the modeler.

4.4.4 Modeling Water Saturation

There are two primary sources of data for saturation-height modeling: core-derived capillary pressure measurements and saturation estimates from log data. Capillary pressure (P_c) is expressed in terms of the interfacial tension between the wetting and non-wetting fluid phases, σ, and the contact angle between the wetting phase and the rock surface, θ, as follows:

$$P_c = \frac{2\sigma\cos}{r}$$

where r is the effective pore radius.

Water saturation in the reservoir should be modeled using a saturation-height function derived from logs and core analysis data. It is *not* recommended to model water saturation using a stochastic petrophysical simulation technique. More often than not, however, a simple layer average constant value is used based on a log-derived average value, and to achieve this, the model becomes a simple layer-cake construction with too many zones. Always check what water saturation you are being given to model: total, effective, initial, or irreducible?

A better alternative is to use a saturation-height relationship derived from logs based on the capillary pressure and the density difference between the two fluids. There are a number of ways to model water saturation based on the physical principle of how a reservoir, initially filled with water, is 'drained' as the hydrocarbons fill the trap displacing the water. In this case, it is necessary to establish the free water level (FWL) at which water saturation is 100% and capillary pressure is zero: this is a datum defined by the physics of the reservoir (Figure 4.22). In clean, homogenous, porous sands the FWL will likely be the same as the hydrocarbon–water contact and water saturation will vary with height above the FWL in a predictable way. In heterogeneous sandstones water saturation will vary as a function of porosity and permeability, making identification of the FWL challenging. Geologists and petrophysicists often describe the increase in water saturation toward a hydrocarbon–water contact as the 'transition zone', but to a reservoir engineer the transition zone is where both oil and water are produced during production.

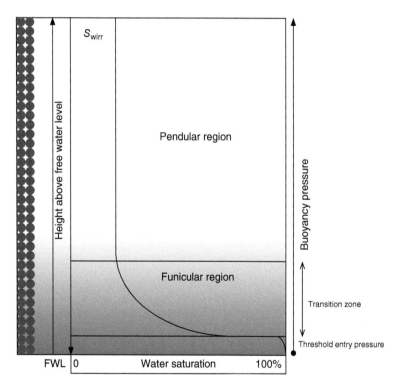

Figure 4.22   Physics of the reservoir: representation of the fluid distribution within an oil reservoir based on the relationship between water saturation, capillary pressure, and the free water level (FWL) datum. *Source:* Cannon. S.J.C. (2016) Petrophysics: a practical guide. John Wiley & Sons, Ltd. © 2016 John Wiley and Sons.

4.4.5 Modeling Net : Gross (NTG)

An arbitrary classification of the reservoir into sand and shale automatically introduces the concept of net and non-net rock in the gross thickness in a well or gross rock volume of a model. In the real world, however, the sand content of the reservoir is variable leading to the need for a property that reflects this variability: hence the concept of a NTG ratio. In classic map-based techniques, NTG is a very important factor when modeling the reservoir. This is because there is insufficient vertical resolution in the 2D model; however, in a 3D grid-based model, the vertical resolution is taken into account in the grid building.

NTG reservoir in a well is usually generated by the petrophysicist based on a cut-off in estimated shale volume (V_{sh}); sometimes a porosity cut-off may also be used independently or in conjunction with shale volume (Figure 4.23). If you want to build a facies model that will give similar results you must agree a facies scheme that either drives the choice of cut-offs or reflects those applied by the petrophysicist: you cannot work in isolation. Never forget that NTG is a ratio, so when combining values from wells or zones it is necessary to compare both the net and the gross thickness.

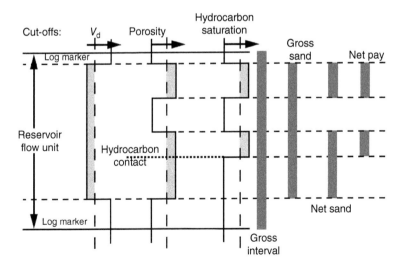

Figure 4.23 Net : gross (NTG) terminology: whichever approach you take be consistent. *Source:* Cannon. S.J.C. (2016) Petrophysics: a practical guide. John Wiley & Sons, Ltd. © John Wiley and Sons.

When building facies-constrained models, target volumes are set for each reservoir property and the background is usually set to non-reservoir, thus automatically introducing a NTG relationship. Use the mantra 'does it add volume or flow' to help define what is net reservoir. The key to understanding NTG and applying it in reservoir modeling is to be consistent in your terminology and definitions across all the disciplines. One important issue to remember is that in blocking the facies data we have introduced a change in the log-derived NTG property due to the upscaling of the data; as a result, the input and the blocked data will not always be consistent.

4.4.6 Modeling Permeability

Permeability is a dynamic, vector property having both value and direction, so anisotropy is commonly modeled both vertically and laterally. Permeability is usually grouped into the same facies classes as porosity but can have a larger variation in values because it is log-normally distributed (a skewed distribution). Vertical permeability is often calculated based on the horizontal permeability; most times a $K_v : K_h$ ratio is used. For each facies group, parameter histograms and cross plots must be made in order to validate the assumptions. For facies with few samples one may have to rely on concepts and assumptions regarding how well such properties are correlated. Permeability is important for flow simulation; time is therefore needed to analyze the controls on distribution and to model it correctly. In the dynamic model, the relative permeability of one fluid in the presence of another is a key input that may be derived from experimental data or be based on empirical models.

Commonly, permeability will be distributed based on the porosity model by either co-kriging or co-simulation; the linear function relationship is not recommended. In each of these cases, a permeability value is selected from the distribution that corresponds to a given porosity in the cell. The difference in the two methods is that in co-simulation the positioning of the cell is randomly chosen.

4.5 Upscaling

Upscaling is finding the single property value that best represents the heterogeneity of a group of cells in the fine-scale model to be used in a coarse-scale simulation grid (Figure 4.24). The challenge is to maintain

Fine-scale

Coarse-scale

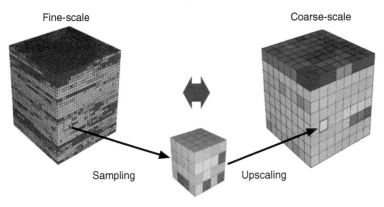

Sampling

Upscaling

"A cynic would describe upscaling as putting incorrect information
into a model to the get right answer" (*Mike King, BP*)

Figure 4.24 Upscaling of reservoir properties is dependent on sampling
method, scale, and region. *Source:* Cannon. S.J.C. (2018) Reservoir Modelling: a
practical guide. John Wiley & Sons, Ltd. © 2018, John Wiley and Sons.
Reproduced with permission of Schlumberger-NExT.

the content of the fine-scale model in the upscaled grid; this especially
true of the different scales of heterogeneity that will impact on production
and recovery of hydrocarbons. Even with the increasing power of
modern computers and parallel processing, it is seldom possible to run
dynamic simulation on a fine-scale, full-field model. When a reservoir
engineer is initializing a dynamic model, they are trying to maintain the
total pore volume of hydrocarbons. To achieve pore volume correspondence, a simple summation method is used; however, when it comes to
flow in the model, different permeability upscaling methods are required,
and these should be based on well test data and production history.

4.6 Model Analysis and Uncertainty

Before building a model, the input data are analyzed for statistical content so that they might be used 'correctly' to populate a model. The
results must also be analyzed as part of the quality control process; the
term 'garbage in-garbage out' has never had more meaning than in
statistical modeling. The input and output statistics should be compared
in terms of mean values and standard deviation; however, if any trends

or conceptual ideas have been applied to control property distribution it is unlikely that the properties will match. There is a common misconception that the well data should be explicitly be recreated through the whole model; this will only be true if a deterministic model is built using well data alone.

Generally, the static model will be used for in-place hydrocarbon volumes estimation or connected drainable volumes. However, using stochastic rather than deterministic methods allows the modeler to build multiple realizations of any given scenario. Scenarios are user-defined and could represent differing conceptual ideas, such as channel orientation or top reservoir structure; these would be deterministic uncertainties. The different realizations are the results of stochastic uncertainty within each scenario. By ranking these multiple realizations, it is possible to identify specific cases representing low, medium, and high outcomes or P90, P50, and P10 results.

4.7   Summary

For a more detailed discussion and review of these topics I refer the reader to my book *Reservoir Modeling: A Practical Guide* published in 2018 (John Wiley & Sons).

Never forget '*all models are wrong, some are useful!*' (Box 1979).

5

Performance, Monitoring, and Forecasting

An important part of reservoir management is to understand the expected reservoir performance at every stage of the development not just at production start-up, but later when looking to enhance recovery through secondary and tertiary stages of extending field life. This task is improved by regular or permanent monitoring of the reservoir. Additionally, estimation of reserves and recovery factors under different producing mechanisms is essential to maximize the potential of the accumulation. Understanding the range of hydrocarbons initially in-place (HIIP) is important for the management, accounting, and reporting of the business.

Between 2007 and 2011, the Society of Petroleum Engineers (SPE), the World Petroleum Council (WPC), and the American Association of Petroleum Geologists (AAPG), together with the Society of Petroleum Evaluation Engineers (SPEE) and Society of Exploration Geophysicists (SEG), developed a standard classification of resources and reserves that ensures that within the Western world, at least, all potential hydrocarbon development projects can be measured by the same yardstick: we will look at the Petroleum Resource Management System (SPE [PRMS] 2018) in Section 5.5.

Reservoir Management: A Practical Guide, First Edition. Steve Cannon.
© 2021 John Wiley & Sons Ltd. Published 2021 by John Wiley & Sons Ltd.

5.1 Natural Drive Mechanisms

Fluid movement in a hydrocarbon reservoir is a function of three forces acting at the microscopic scale: capillary forces, viscous forces, and gravity. These natural forces are reflected in variations in reservoir pressure, production rates, fluid ratios, aquifer influx, and gas cap expansion. Each of these are further influenced by geological characteristics, rock and fluid properties, fluid flow mechanisms, and production facilities. Essentially, energy in the reservoir comes from the liberation of gas in solution, influx of aquifer water, the expansion of reservoir rock and compression of pore volume, the expansion of original reservoir fluids (gas, interstitial water, and oil), and, finally, gravity where the reservoir is of sufficient thickness (Figure 5.1).

The vertical line in Figure 5.2 shows the ideal isothermal depletion for the different types of reservoir fluids:

- The *crude oil* envelope shows that the initial pressure and temperature corresponds with an undersaturated oil. As the pressure drops below the bubble point some gas is liberated from solution.
- The phase behavior of a *volatile oil* is similar except that a larger volume of solution gas is liberated. Because this is related to the

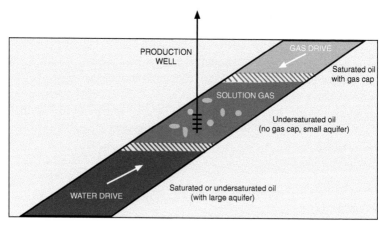

Figure 5.1 Schematic of the primary drive mechanisms found in an oil or gas field describing the different initial fluid conditions.

intermediate components in the fluid tending to become gas rather than a liquid, they are referred to as 'high shrinkage oils'.

- In a *gas condensate* the reservoir temperature is higher than the critical temperature and a two-phase region exists such that as the pressure drops the dew point is crossed. When the gas condenses into a liquid

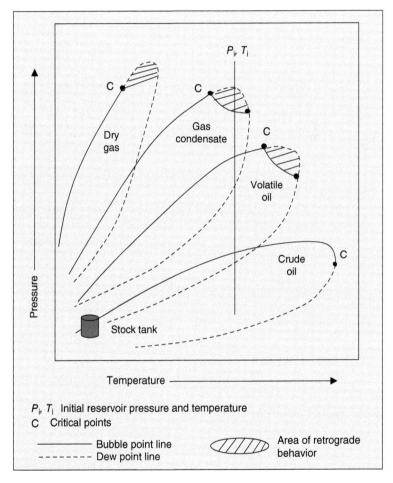

Figure 5.2 The phase behavior of the main types of hydrocarbon reservoirs. The vertical line shows the expected changes in pressure and temperature that occur as a consequence of hydrocarbon production.

Table 5.1 Notional oil and gas recovery factors based on primary recovery mechanisms.

Drive mechanism	Average oil/gas recovery factors (% of HIIP)	
	Range	Average
Solution gas drive	5–30	15
Gas cap drive	15–50	30
Water drive	30–60	40
Gravity-drainage drive	16–85	50
Gas expansion drive	70–90	80
Water drive (gas field)	35–65	50

under isothermal conditions rather than vaporizing it is known as a retrograde condensate.

- For a *dry gas* to exist the reservoir temperature is always higher than the critical temperature, even during isothermal depletion, and no liquid drop out occurs. These gases are generally formed of the lightest hydrocarbons, methane, and ethane.
- A *wet gas* is very similar to a dry gas except that when the pressure–temperature conditions occur in the two-phase region some liquid can be expected to condense at the surface.

Each drive mechanism can respond differently to the type of hydrocarbon in the reservoir resulting in the range of recovery estimates (Table 5.1).

5.1.1 Fluid Expansion

In this simple case, reservoir fluids expand as pressure declines; in effect the fluids displace themselves. This is the dominant process in gas and gas condensate reservoirs and can lead to recovery factors in the order of 80%. In the case of undersaturated oil reservoirs, there is a rapid decline in pressure, but the gas–oil ratio (GOR) remains constant, resulting in very low recovery, less than 5%, because oil compressibility is usually

FLUID EXPANSION

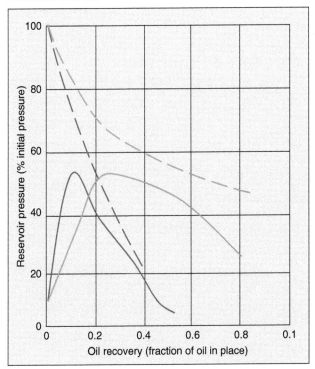

Figure 5.3 Typical response of oil (red) and gas (green) to the decline in reservoir pressure leading to a primary recovery factor of 0.4 and 0.8 respectively.

very small, especially with heavier oils. This is the least efficient drive mechanism (Figure 5.3).

5.1.2 Solution Gas Drive

As pressure declines during production, gas is liberated that expands and displaces the oil. However, gas will only flow after reaching a critical saturation (2–10%). Once the critical saturation is reached, gas flows at a velocity proportional to the saturation: as the pressure drops more gas is released in a domino effect, further reducing pressure leading to rapid depletion of the reservoir. There may be a rapid increase in GOR and decrease in oil rate, with little associated water production. The ultimate recovery in this case is ~10–35% of oil initially in-place. The gas–oil relative permeability is the most important reservoir parameter to describe and understand when modeling this form of drive mechanism. In some cases, especially where the reservoir interval is thick (10's–100's m),

SOLUTION GAS DRIVE

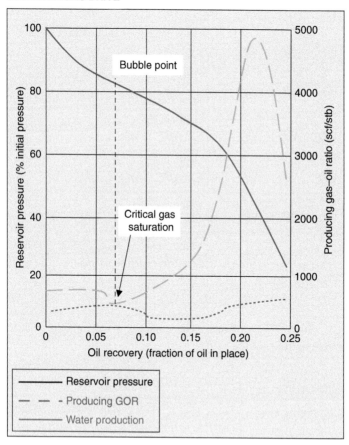

Figure 5.4 Reservoir fluid response under solution gas drive: as the reservoir pressure declines due to production, the gas–oil ratio rises rapidly after the bubble point has been exceeded. Oil production declines rapidly and ultimate recovery is reduced unless pressure maintenance can be introduced.

gravity segregation can cause a secondary gas cap to form and greater ultimate recovery. Wells drilled updip in the reservoir may have higher gas production, which requires management (Figure 5.4).

5.1.3   Aquifer Drive

Where the reservoir is connected to an active natural aquifer, water expands into the reservoir displacing oil. This may be represented by a bottom or flank (peripheral) drive. If the water influx is steady and uniform a piston-like drive is observed and ultimate recovery improved. Where there is significant geological heterogeneity there may less effective displacement of oil resulting in water fingering. Water coning may be observed if the wells are produced at too high a rate. The size of aquifer is estimated by a simple ratio of the radius of aquifer versus the radius of the reservoir ($r_D = r_A/r_R$) and a figure of 50 represents a strong drive: an infinite acting aquifer has no apparent boundaries. The reservoir properties of the aquifer should also be considered separate from the reservoir itself, especially in terms of permeability. Higher aquifer permeability increases the effectiveness of the drive; however, permeability in the aquifer can be reduced due to continuing diagenesis after hydrocarbon migration and trapping. During production, GOR remains the same as long as reservoir pressure is above the bubble point for the fluid. Recovery can be highly variable depending on the local conditions; however, an expected recovery factor of 30–80% is not unreasonable. The water–oil relative permeability is an important reservoir parameter at the microscopic level (Figure 5.5).

5.1.4   Gas Cap Drive

In the case of a gas cap drive, during production free gas expands into the oil leg displacing oil toward downdip wells. Where a primary gas cap exists at discovery, the oil phase is saturated and the pressure at the gas–oil contact is equal to the saturation pressure. Therefore, any depletion of the reservoir releases gas from solution, such that both drive mechanisms are active at the same time; which is more effective depends on the relative size of the gas cap. In the case of a secondary gas cap forming after a period of production, the effects of gravity segregation must be operating for gas to subsequently migrate updip in the structure. For this to take effect there must be a high vertical permeability, favorable

AQUIFER DRIVE

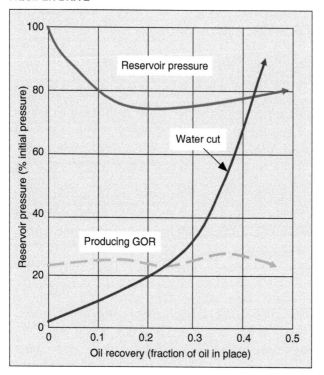

Figure 5.5 Recovery profile of a reservoir with an active aquifer drive: the water ingress supports the reservoir pressure but at the expense of increasing water production. Ultimate recovery can be improved as well.

mobility, low flow velocity, and thick or dipping strata. A recovery factor of 30–70% is achievable, dependent on the dimension of the gas cap, effectiveness of gravity segregation, and efficiency of the gas displacement process (Figure 5.6).

5.1.5 Compaction Drive

This drive mechanism is related to a pore volume decrease due to fluid withdrawal: pore compressibility must be small compared with fluid compressibility. In effect, we are looking at the difference between the overburden pressure and the formation fluid pressure ($P_{eff} = P_{tot} - P_{fluid}$). This effect is probably active in most reservoirs but is often ignored as

GAS CAP DRIVE

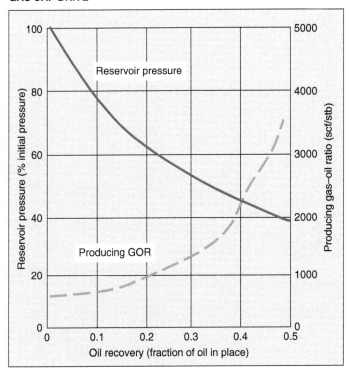

Figure 5.6 With a gas cap drive, as reservoir pressure declines the producing gas–oil ratio increases.

being minimal compared with the primary drive mechanism. Significant exceptions exist in the literature such as fields in the Maracaibo Basin, the Ekofisk chalk reservoir, and the giant Groningen gas field. In some cases, rock compressibility may account for up to 50% of reservoir energy; however, the production behavior is difficult to define or predict: such fields should be treated like they have a natural water drive without the problems! Material balance techniques may be used to resolve some of the issues in reservoir management (Figure 5.7).

5.1.6 Combination Drive

In reality, of course a reservoir under production may show each of these natural drive mechanisms during its lifetime; in essence a combination drive.

COMPACTION DRIVE

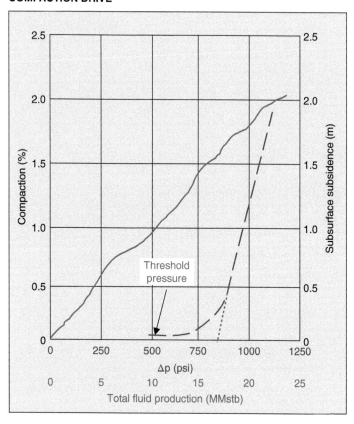

Figure 5.7 Under a compaction drive mechanism, when the threshold pressure is exceeded the degree of pore volume compaction maintains fluid production but results in surface subsidence.

5.2 Reservoir Monitoring

Permanent reservoir monitoring is one the most important advances in reservoir management over the last 20 years. Smart wells and intelligent fields are terms that have entered the vocabulary of petroleum engineers, but perhaps they need some definition (Ezekwe 2012):

- *Smart wells* are equipped with electronic sensors that enable remote monitoring, control, and transmission of data from multiple zones, thus giving the opportunity to optimize production, improve well management, and reduce intervention costs.

- *Intelligent fields* have centralized monitoring, control, and management of individual wells with a view to implementing and executing predefined reservoir management strategies.

Various downhole sensors (DHS) are installed to monitor, measure, and transmit data from the well on fluid flow, fluid properties, pressure, and temperature as well as other well performance information. These include permanent and distributed pressure and temperature sensors, sand production sensors, microseismic sensors, and single phase and multiphase flow meters.

Permanent downhole gauges (PDG) are the most widely used DHS with more than 20 000 units installed in every setting worldwide. PDG can be used to measure static and flowing pressure in the well, reducing the need for shut-in tests, and to monitor inflow performance. Optical flowmeters can be configured for use as single or multiphase measurement tools, providing long-term stable and reliable results. These gauges can offer full through bore access to the well with no pressure drop along the length of the tool. Multiphase meters can also combine a physical Venturi measurement of mass flow rate with a dual energy gamma ray meter for the measurement of individual gas, oil, and water fractions. These sophisticated meters can replace conventional separator systems; however, there may be issues with mechanically pumped wells, as the flow rates tend to be erratic. Downhole multiphase meters can reduce the need to install test lines, separators, and manifolds, reducing capital and operating expenditure on offshore installations as well as the need for space.

Downhole control devices (DCD) are used to regulate specific intervals of a well to achieve zonal control of fluids without the need for physical intervention. The two main types of control devices are interval control valves (ICV) and inflow control devices (ICD). These devices come into their own in multilateral wells or reservoir layers where there are significant pressure differences or flow variation. The main benefit of an ICD is to actively manage uniform flow distribution along a horizontal wellbore. Additionally, DCD have revolutionized the ability to conserve reservoir energy especially by reducing the production of free gas, achieving one of the main objectives of reservoir management, energy conservation.

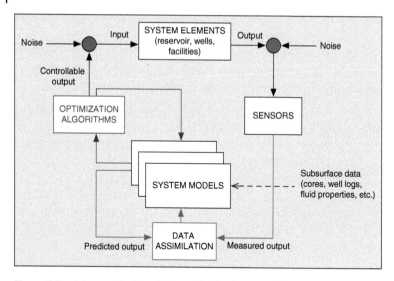

Figure 5.8 Advances in downhole monitoring have resulted in advanced closed-loop systems in which the real-time data are used to optimize production from individual wells and a field as a whole.

DHS have increased our ability to collect large quantities of high-quality data from wells than previously available from surface devices. These in turn have allowed for that data to be transmitted, stored, analyzed, integrated, and used for near real-time reservoir management (Boe et al. 2000). This has created an advanced closed-loop system in which the real-time data are used to optimize production from individual wells and a field as a whole (Figure 5.8).

5.3 Production System

From the reservoir level to the surface and at every step in the production of hydrocarbons there is a reduction in pressure. The reservoir provides the source of the fluids and the initial energy, as well as the media of storage and flow for the fluids. The well bore is the conduit for access to the reservoir, usually via a series of tubulars that house the subsurface control equipment. At the wellhead the surface production equipment controls flow rates and pressures as well as separates, treats, and stores the produced fluids. At this stage, artificial lift may be applied by pumps or gas compression. Downhole pumps are available in extended reach wells, but successful downhole compression is still a work in progress. Each well is part of a larger integrated system of many wells and surface facilities through which the reservoir is managed: a single well cannot be analyzed as a stand-alone unit but part of the whole. This best achieved by a simulation model where each well can be operated individually, and the impact studied in terms of the producing area or the whole field.

5.3.1 Well Completion Methods

The choice of a well completion can have an impact on ultimate production from a well, both positively and negatively: each method has advantages and disadvantages. Different solutions are needed for single or multiple reservoirs, for deviated and horizontal wells, for homogenous and heterogeneous reservoirs: in other words, choosing the best well completion can have many drivers. Completion of a production well may be done by the drilling rig at the end of the drilling operation or the well may be suspended and a workover or completion rig utilized, often at a lower cost.

Well completion operations provide the means of communication between the reservoir and the surface and include perforating, sand control installation, setting a production packer, running a production tubing string and its components, installation of a subsurface safety valve (SSSV), and the Christmas tree equipment. There are four main well completion techniques that may be deployed in different circumstances: open hole, liner, cased hole, or slim hole. The open hole completion is also called a barefoot completion because there is no protection for the reservoir.

5.3.2 Open Hole Completion

The main advantage of an open hole completion is that it eliminates the cost of casing and perforation and gives the maximum well bore/reservoir contact; however, it is difficult to control excessive water or gas production and can only really be considered for well consolidated reservoirs. On a positive note, it is easy to convert to a screen or perforated completion at a later stage: it is also possible to deepen the well later (Figure 5.9).

5.3.3 Liner Completions

There are two types of liner completions: cemented liners and slotted screens or perforated liners. Screens or liners have the advantage as they do not require perforating and the screen can be designed to provide instant sand control in poorly consolidated reservoirs: a slotted liner does not provide sand control. Borehole stability is improved with both types of completion, but again excessive water or gas influx is difficult to control and it not easy to selectively stimulate producing intervals. A higher 'skin' due to the presence of mudcake can be expected as well as the presence of a barrier with the reservoir made by the liner in the first place, both of which will reduce natural flow (Figure 5.9).

5.3.4 Cased Hole Completions

In many cases, this is the most effective type of completion as it allows water and gas production to be controlled, as well as selective stimulation and multiple completions, and is readily adaptable for sand control techniques. However, perforating cost can be significant as well as the cost of the extra casing, and there is a risk of causing a completion skin, further reducing flow from the reservoir (Figure 5.9).

5.3.5 Perforated Liner Completions

Use of a perforated liner has all the same advantages and disadvantages as a perforated cased hole completion including the ability to deepen the borehole at a later stage; however, cementing the liner in place is more challenging (Figure 5.9).

5.3.6 Slim Hole Completions

The greatest advantage of a slim hole completion is the lower cost; however, the limited workover capability due to small borehole diameter and the reduced stimulation rates are strong negatives (Figure 5.9).

5.3.7 Designer Wells

The advent of extended reach and horizontal wells has resulted in numerous inventive well completions, including single and multiple producing zones. Initially, these wells were completed open hole but there were problems with sand control and excess gas or water production especially at the heel of a horizontal well. In a deviated well, up to ~65°, the standard oilfield equipment is usable, but above this curvature running casing can become an issue as can the installation of artificial lift equipment. Present day horizontal well completions require new or

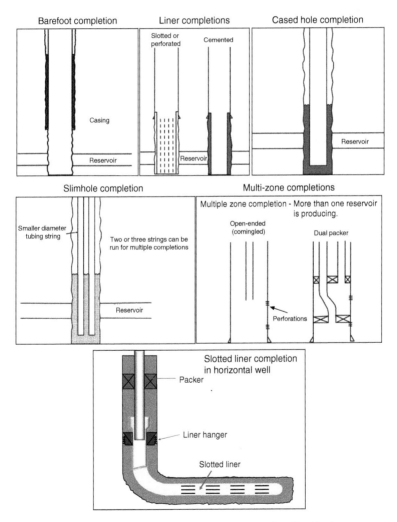

Figure 5.9 Typical types of well bore completions as described in the text.

modified equipment and methods to be successful. When a horizontal drainhole is close to either the gas–oil or oil–water contact, using a cemented and perforated casing or liner is the best solution, especially as it possible to orient the perforation to minimize coning or cusping of the non-producing fluid. An effective well and its completion is one that maximizes the return of the life of the well in terms of capital cost and operational expense (Figure 5.9).

5.4 Resource and Reserves Estimation

We estimate resources so we can measure the commerciality of a discovery or development (Figure 5.10). Different organizations have different processes to estimate the volume of hydrocarbons and different hurdles to jump before sanctioning a project as commercial; we are providing our best estimates of HIIP so that these decisions can be made successfully. Companies are in the business of making money and satisfying their stakeholders and investors; increasingly, the stakeholders include the national governments where they operate. Operators need to manage their assets effectively to conduct business and to manage expectations, so the value of a portfolio of assets must be estimated for the timing of future development. Additionally, most publicly quoted international oil and gas companies have a legal requirement to disclose the value of the resources and reserves to stock exchanges so that investors can make informed decisions: stock exchanges do not make a recommendation to potential investors but ensure that a company is a viable trading concern. It is worth reiterating that we *estimate* reserves and resources and that the process is not an exact science.

5.4.1 Volumetric Estimation

Before understanding reservoir performance or forecasting the likely ultimate production, it is necessary to know the size of the prize: how much hydrocarbon is initially in-place. There are two primary methods of estimating the volume of hydrocarbons:

- *Direct measurement* using volumetric calculation, which is a static measurement based on the area of the field, the thickness of the reservoir, the net rock volume, porosity and hydrocarbon saturation,

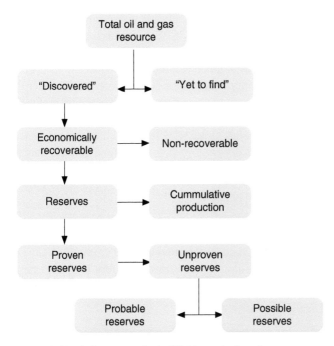

Figure 5.10 Industry standard oilfield terminology for resources and reserves. *Source:* Cannon. S.J.C. (2018) Reservoir Modelling: a practical guide. John Wiley & Sons, Ltd. © 2018, John Wiley & Sons.

together with an expansion factor. The accuracy of the result is a function of the measurement of each of these parameters: more data points and better measurement methods can improve the accuracy.

- *Indirect measurement* using performance analysis, which is a dynamic method based on the initial reservoir pressure, temperature, and analysis of the fluids recovered. During the production of hydrocarbons these properties are repeatedly measured and recorded and used to estimate ultimate recovery.

Knowledge from offset analogous wells/fields is often used where there is sufficient information to build a robust input data set.

5.4.1.1 Direct Measurement
The simplest way to calculate the volume of hydrocarbons is a deterministic estimate based on the gross rock volume (GRV) of the reservoir,

the net: gross (NTG) ratio, and the porosity and water saturation (hydro-carbon saturation $S_h = 1 - S_w$). The GRV is calculated from a top structure map and the free water level (FWL) either as a series of area depth maps or from a geocellular model. Mapped volumes are measured with a plan-imeter and isochore maps are summed for different areas using macros. We have learned from Chapter 4 how to generate a 3D geocellular model. Porosity is estimated from well data and distributed by the chosen method, deterministic or stochastic; NTG is determined from a facies model or well data; the water saturation is distributed using a height above FWLClick or tap here to enter text. relationship and the volume factor calculated from laboratory pressure–volume–temperature (PVT) experiments or offset data (Figure 5.11).

Often the best way to develop a volumetric model is to refine an exist-ing deterministic estimate of hydrocarbon pore volume or HIIP. This may then form a base case model against which further estimates can be compared. In this way the impact of different contributing parameters and their uncertainty can be assessed. The oil and gas industry has a

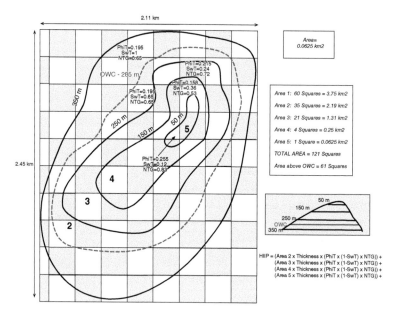

Figure 5.11 Use of an area depth map is the oldest form of volumetric estimation.

consistent approach to describing the resources in a field or a basin (Figure 5.12.).

A common *deterministic* approach to defining reserves is in terms of proven, probable, and possible categories; the *probabilistic* approach is to refer to P90, P50, and P10 deciles on the cumulative distribution curve. These terms are quite specific:

- *Proven*: lowest risk, reasonable certainty of production under existing economic conditions; a 90% probability of production and similar to

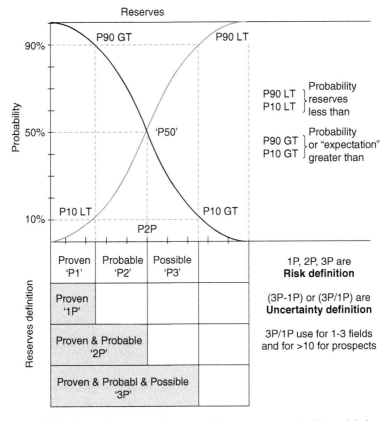

Figure 5.12   A standard approach to describing resources using deterministic and stochastic terminology. *Source:* Cannon. S.J.C. (2018) Reservoir Modelling: a practical guide. John Wiley & Sons, Ltd. © 2018, John Wiley & Sons.

the P90 probabilistic definition of there being a 90% chance of producing the booked amount.

- *Probable*: more likely to be produced than not; at least a 50% probability of production and similar to P50 such that a business decision can
  be made on the 'most likely' or 'expected' outcome.
- *Possible*: more speculative, a higher degree of risk with at least a 10%
  chance of being developed; an upside case for the maximum return on
  the investment.

It is also possible to use both types of classification to compare or rank
prospects, projects, and investment opportunities. The terms 1P, 2P, and
3P define the risk associated with a project, while comparisons of 1P and
3P can define the degree of uncertainty: the ratio of 3P : 1P should be
between 1 and 3 for fields and greater than 10 for prospects. A spread of
around 3–5 should be expected for a field that is pending development.

5.4.1.2 Indirect Measurements

Indirect or dynamic methods of volumetric estimation include analogy,
material balance, decline analysis, and reservoir simulation: together
they are all performance-based methods of analysis. The first two of
these analytical methods are often described as classical reservoir engineering and are dependent on reliable measurement of the reservoir
pressure, temperature, and fluid production, as well as fluid injection or
aquifer movement. Reservoir simulation is the only method that can give
the engineer hydrocarbon production profiles under differing development scenarios (Figure 5.13).

Using *analogy* is a simple way of comparing a field under development
with a similar accumulation with an existing production history. In its
most basic form, analogy will give some indication of both individual
well production and ultimate recovery from a similar field. In an onshore
situation it is common to see maps constructed equal *estimated ultimate
recovery* (EUR) (Figure 5.14); some additional geological information
adds to the value of the analysis. Another way to look at data from an
analogous field is a plot of production profile against time and well
count; this is essentially a type curve of weighted average of observed
actual performance.

Decline curve analysis is a very subjective way of looking at the decline
in hydrocarbon production over time (Figure 5.15). Decline analysis

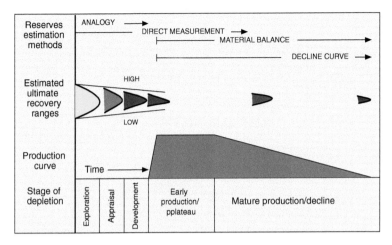

Figure 5.13   The different indirect methods of reserve estimation using analogy and production pressure data to define material balance and decline curve analysis.

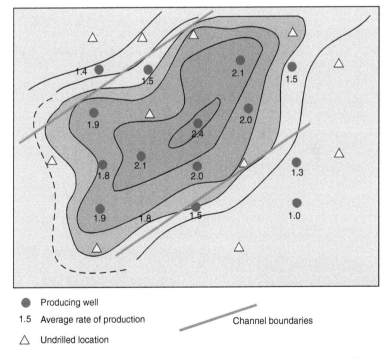

- ● Producing well
- 1.5   Average rate of production
- △   Undrilled location
- ─── Channel boundaries

Figure 5.14   An example of analogy based on past well production data. The map shows well locations and average production rate to identify sweet spots in the reservoir.

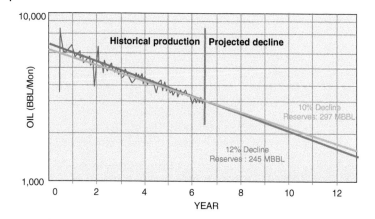

Figure 5.15   Decline curve analysis based on historical production. The rate of decline can be very subjective: should you take the total production data or just the last 1–2 years?

extrapolates the historical performance trend to an economic production limit or cut-off, thus forecasting the ultimate recovery expected. The method plots the production rate against time and records the decline as cumulative production increases. Theoretically, the techniques should only be applied on a single well, but extrapolation for groups of wells will usually provide a reasonable estimation of ultimate recovery. It is assumed that all the factors affecting production remain constant; any changes to the production methods will have an impact on the decline curve. The value in a decline curve analysis is that it is much simpler to initiate than a full field reservoir simulation.

Common decline curves used to analyze oil reservoirs include the logarithm of total fluid production rate against time, production rate against cumulative production, and the logarithm of water-cut or oil-cut against cumulative production. When the shape of the curve alters, the cause should be found before further analysis is attempted. The question arises whether the estimate should be made over the whole period of production or just over the more recent past. The difference in the rate of decline can impact on both reserve estimates and future production.

Material balance is typically used for volumetric gas field prediction, especially where there is a long-term contract for a fixed volume of gas production. The basic assumption is that rock and fluid properties are constant, that is, a 'tank' model, and that production and injection occur

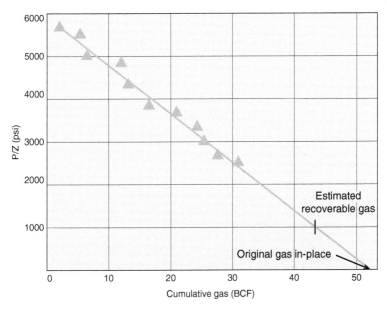

Figure 5.16   Material balance is regularly applied in gas fields under a fixed contract to estimate the recoverable gas. The P/Z factor relies on the average reservoir pressure and the gas deviation from an ideal gas at the current temperature, together with the cumulative production.

at single but separate points. It is, however, notoriously difficult to validate the prediction where there is an active aquifer that generates additional reservoir energy. The plot produced is of cumulative gas production against the pressure decrease as a function of gas compressibility. If the original oil in-place is known as well as the gas cap size, aquifer strength and size are known, material balance can be used to predict future performance (Figure 5.16.).

Material balance is an expression of the law of energy conservation and its value in the estimation of HIIP is that calculation is independent of any of the variables used in a static volumetric calculation. The effective use of material balance is predicted on the assumption that there is pressure equilibrium throughout the reservoir at each time-step, and that there is reliable production and representative PVT data. Material balance should be considered as a precursor to a dynamic simulation solution as most of the inputs needed for a simulation study can be investigated through a material balance exercise.

Reservoir simulation is fundamental tool available to the reservoir engineer when it comes to understanding flow in the reservoir and predicting rates of production and ultimate recovery in all development scenarios. Engineers use numerical simulation because of the applicability of the tool to solve every problem they may encounter. Today's simulation packages are relatively interactive and have sufficient pre- and post-processing functionality to maximize the use of the results. Dynamic reservoir models are used to develop the management plan and to monitor performance and evaluate the results of production (Figure 5.17).

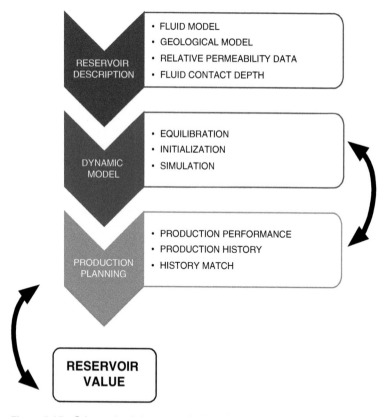

Figure 5.17 Schematic of the reservoir simulation process from model build, initialization, and analysis.

Numerical simulation is essentially based on the principles of material balance, but where reservoir heterogeneity and direction of flow can be taken into account. Multiple producer and injector well locations can be incorporated in the model, and under different operating conditions. Well control is optimized such that completions can be opened and closed, and both well rates and pressure can be managed under different situations. In effect, a reservoir can be divided into smaller tanks both vertically and laterally, and computations using material balance and fluid flow equations can take place at regular time-steps through the life of the field.

History matching of a simulation model consists of adjusting the reservoir variables until past production and pressure parameters are met both for the field as a whole and ideally for each well in the model. If this can be achieved, then a forward prediction may be modeled by extending the time parameter. History matches are not unique: a number of different reservoir descriptions can produce the same history match but different forecasts. A step-by-step history matching process is recommended:

1) Initialization of the model with the known properties and conditions.
2) Pressure matching of producing wells and injectors and adjustment of the parameters that impact on the original HIIP estimates.
3) Saturation matching of oil, water, and gas by adjustment of relative permeability curves, rock types, and hydrocarbon contacts.
4) Well pressure matching by modifying the production indices (PI).

Predicting the future performance of a field or a group of wells under existing operating conditions or under a revised development plan is the ultimate objective of a simulation study.

Reservoir modeling and simulation are both much abused these days, especially where computer software and hardware have created too much reliance on these techniques. This has resulted in unrealistic expectations and the pushing of input data and concepts too far, but the greatest risk is the unwarranted dependence on the answers generated. By going through the process of building a database it helps in clarifying the reasons for the study, identifying the requirement for additional data, and evaluating the quality and quantity of the data. So don't start a simulation study without clear objectives; understand those parameters that have the greatest impact on the study outcome and take care with prediction results that have not been validated in the history match.

5.5 Petroleum Resources Management System (PRMS)

In 2018, SPE, World Petroleum Council (WPC), American Association of Petroleum Geologists (AAPG), and SPEE jointly released new guidelines for the application of the PRMS. PRMS is a means of defining hydrocarbon resources and reserves designed to provide a 'consistent approach in estimation of quantities, evaluating development projects and presenting the results within a comprehensive classification framework' (SPE [PRMS] 2018). The definitions and guidelines are designed to provide a common reference for the international petroleum industry, including national reporting and regulatory disclosure agencies, and to support petroleum project and portfolio management requirements. They are intended to improve clarity in global communications regarding petroleum resources. One particular advantage of the system is that it can be applied to conventional and unconventional volumes of hydrocarbons.

The system is designed to differentiate between resources, all naturally occurring hydrocarbons in a field or license area or a basin, and reserves, which are quantities of hydrocarbon that are expected to be commercially recoverable by a sanctioned project. Reserves must satisfy four criteria: they must be discovered, recoverable, commercial, and remaining to be produced from a given date. Both categories may be classified as proved (P1), probable (P2), or possible (P3), with a range of uncertainty estimated either deterministically (low–best–high) or stochastically (P90–P50–P10) (Figure 5.18). Tables with the definitions of each class of resource or reserves are presented in Tables 5.2–5.4.

To be included in the Reserves class, a project must be sufficiently defined to establish its commercial viability. There must be a reasonable expectation that all required internal and external approvals will be forthcoming, and there is evidence of firm intention to proceed with

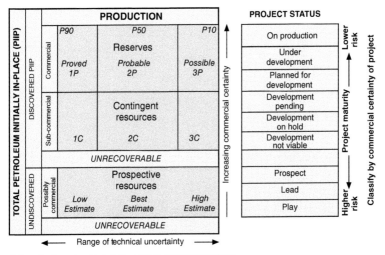

Catagories based on technical certainty of sales quantities

Figure 5.18   Petroleum Reserves Management System (PRMS) (SPE [PRMS] 2018) terminology for resources and reserves. *Source:* Society of Petroleum Engineers. Cannon. S.J.C. (2018) Reservoir Modelling: a practical guide. John Wiley & Sons, Ltd. © 2018, John Wiley & Sons..

development within a reasonable time frame. There must be a high confidence in the commercial producibility of the reservoir as supported by actual production or formation tests. In certain cases, Reserves may be assigned on the basis of well logs and/or core analysis that indicate that the subject reservoir is hydrocarbon-bearing and is analogous to reservoirs in the same area that are producing or have demonstrated the ability to produce on formation tests.

A reasonable time frame for the initiation of development depends on the specific circumstances and varies according to the scope of the project. While five years is recommended as a benchmark, a longer time frame could be applied where, for example, development of economic projects are deferred at the option of the producer for, among other things, market-related reasons, or to meet contractual or strategic objectives. In all cases, the justification for classification as Reserves should be clearly documented (Tables 5.2–5.4).

Table 5.2 Reserves definitions as recommended in the petroleum resources management system (PRMS).

Class/subclass	Definition	Guidelines
Reserves	Reserves are those quantities of petroleum anticipated to be commercially recoverable by application of development projects to known accumulations from a given date forward under defined conditions.	Reserves must satisfy four criteria: they must be discovered, recoverable, commercial, and remaining based on the development project(s) applied. Reserves are further subdivided in accordance with the level of certainty associated with the estimates and may be subclassified based on project maturity and/or characterized by their development and production status.
On production	The development project is currently producing and selling petroleum to market.	The key criterion is that the project is receiving income from sales, rather than the approved development project necessarily being complete. This is the point at which the project 'chance of commerciality' can be said to be 100%.
Approved for development	All necessary approvals have been obtained, capital funds have been committed, and implementation of the development project is under way.	At this point, it must be certain that the development project is going ahead. The project must not be subject to any contingencies such as outstanding regulatory approvals or sales contracts. Forecast capital expenditures should be included in the reporting entity's current or following year's approved budget.
Justified for development	Implementation of the development project is justified on the basis of reasonable forecast commercial conditions at the time of reporting, and there are reasonable expectations that all necessary approvals/contracts will be obtained.	In order to move to this level of project maturity, and hence have reserves associated with it, the development project must be commercially viable at the time of reporting, based on the reporting entity's assumptions of future prices, costs, etc. ('forecast case') and the specific circumstances of the project. There should be a development plan in sufficient detail to support the assessment of commerciality and a reasonable expectation that the project has reached a level of technical and commercial maturity sufficient to justify proceeding with development at that point in time.

Table 5.3 Contingent resources definition as recommended in the petroleum resources management system (PRMS).

Class/subclass	Definition	Guidelines
Contingent resources	Those quantities of petroleum estimated, as of a given date, to be potentially recoverable from known accumulations by application of development projects, but which are not currently considered to be commercially recoverable due to one or more contingencies.	Contingent resources may include, for example, projects for which there are currently no viable markets, or where commercial recovery is dependent on technology under development, or where evaluation of the accumulation is insufficient to clearly assess commerciality. Contingent resources are further categorized in accordance with the level of certainty associated with the estimates and may be subclassified based on project maturity and/or characterized by their economic status.
Development pending	A discovered accumulation where project activities are ongoing to justify commercial development in the foreseeable future.	The project is seen to have reasonable potential for eventual commercial development, to the extent that further data acquisition (e.g. drilling, seismic data) and/or evaluations are currently ongoing with a view to confirming that the project is commercially viable and providing the basis for selection of an appropriate development plan.
Development on hold	Discovered accumulations where project activities are on hold and/or where justification as a commercial development may be subject to significant delay.	The project is seen to have potential for eventual commercial development, but further appraisal/evaluation activities are on hold pending the removal of significant contingencies external to the project, or substantial further appraisal/evaluation activities are required to clarify the potential for eventual commercial development.
Development not viable	A discovered accumulation for which there are no current plans to develop or to acquire additional data at the time due to limited production potential.	The project is not seen to have potential for eventual commercial development at the time of reporting, but the theoretically recoverable quantities are recorded so that the potential opportunity will be recognized in the event of a major change in technology or commercial conditions.

Table 5.4 Prospective resources definition as recommended in the petroleum resources management system (PRMS).

Class/subclass	Definition	Guidelines
Prospective resources	Those quantities of petroleum that are estimated, as of a given date, to be potentially recoverable from undiscovered accumulations.	Potential accumulations are evaluated according to their chance of discovery and, assuming a discovery, the estimated quantities that would be recoverable under defined development projects. It is recognized that the development programs will be of significantly less detail and depend more heavily on analog developments in the earlier phases of exploration.
Prospect	A project associated with a potential accumulation that is sufficiently well defined to represent a viable drilling target.	Project activities are focused on assessing the chance of discovery and, assuming discovery, the range of potential recoverable quantities under a commercial development program.
Lead	A project associated with a potential accumulation that is currently poorly defined and requires more data acquisition and/or evaluation in order to be classified as a prospect.	Project activities are focused on acquiring additional data and/or undertaking further evaluation designed to confirm whether or not the lead can be matured into a prospect. Such evaluation includes the assessment of the chance of discovery and, assuming discovery, the range of potential recovery under feasible development scenarios.
Play	A project associated with a prospective trend of potential prospects, but which requires more data acquisition and/or evaluation in order to define specific leads or prospects.	Project activities are focused on acquiring additional data and/or undertaking further evaluation designed to define specific leads or prospects for more detailed analysis of their chance of discovery and, assuming discovery, the range of potential recovery under hypothetical development scenarios.

5.6 Summary

The importance of thorough reservoir performance monitoring cannot be stressed enough. Where fields have been in production for decades the data may not be in digital formats accessible to the modern computer savvy engineers, but it doesn't take long to transcribe the data to a spreadsheet and it also means that the practitioner gets an understanding of how a field may have performed in the past.

6

Improving Hydrocarbon Recovery

Improving hydrocarbon recovery should be at the heart of any integrated reservoir management plan: there should be sufficient flexibility in the plan to allow for any serendipitous opportunities as well. There are three main physics-based reasons that lead to low hydrocarbon recovery after primary and secondary methods have been applied: high oil viscosity, interfacial forces, and reservoir heterogeneity.

Primary recovery factors for mature oil fields around the world can be as little as 5% and up to around 40%: in contrast, typical gas fields show a recovery factor in the order of 80–90% (Table 5.1). If it were possible to significantly increase the recovery factor from oilfields to say 60% it would make a huge difference in global supplies of energy. To achieve this requires better reservoir characterization, improved monitoring of production parameters, and a change in stakeholder culture to plan for the long-term. Only one company, when it was known as Statoil, now Equinor, had a declared aim of achieving 50% plus recovery from all of its assets: hopefully this is still the case.

The production life cycle of an oilfield is characterized by three main stages: buildup, plateau, and decline (Figure 6.1). One of the aims of successful reservoir management is to sustain the maximum production levels during the primary and secondary phases of recovery. To achieve this requires an understanding of the recovery mechanisms at work and the ability to control them effectively. The period of primary recovery is generally short and the recovery factor does not exceed 20% in most cases. For secondary recovery, relying on either natural or artificial water

Reservoir Management: A Practical Guide, First Edition. Steve Cannon.
© 2021 John Wiley & Sons Ltd. Published 2021 by John Wiley & Sons Ltd.

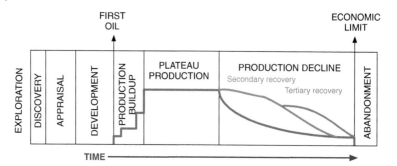

Figure 6.1   Oilfield life cycle from discovery to abandonment with a typical primary production profile in red; secondary and tertiary recovery phases also shown. The application of secondary recovery is usually planned to extend the plateau period.

or gas injection, the incremental recovery ranges from 15–25%. Globally, the overall recovery factors for combined primary and secondary recovery range between 35 and 45%.

To maximize hydrocarbon recovery, it is critical first to have a clear understanding of the static properties and dynamic behavior of the hydrocarbon system on various scales, ranging from the pore scale to the reservoir scale (Figure 6.2). The selection of the type, number, and placement of wells to achieve optimum reservoir drainage requires detailed knowledge of reservoir geology (on a scale of hundreds to thousands of meters). Geological models, however, suffer from uncertainties that inhibit simple deterministic predictions of their flow behavior. Complex geological reservoirs (layered, compartmented, etc.) have been produced using deviated, horizontal, and multilateral wells, and progress in both geophysical imaging and geological modeling will enable complex geological architectures to be exploited more efficiently.

All reservoirs lose energy as hydrocarbons are produced, therefore, to improve recovery usually means introducing a new source of energy into the reservoir or to increase the flow capacity of the wells. These methods generally fall into two categories: thermal and nonthermal recovery methods. Many of the methods that will be discussed are routine reservoir management tools, such as pumping, water injection, or hydraulic fracking and might be called secondary; others involve highly sophisticated chemical or biological additives and are clearly tertiary methods (Table 6.1).

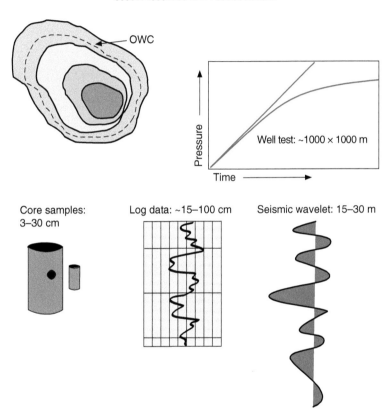

Figure 6.2 Scales of measurement from core data through log, seismic, and well test to demonstrate the several orders of magnitude difference between the various sources of data. *Source:* Cannon. S.J.C. (2018) Reservoir Modelling: a practical guide John Wiley & Sons, Ltd. © 2018, John Wiley & Sons.

6.1 Primary Recovery

Given the wide range in hydrocarbon properties (density, viscosity, PVT properties, etc.), the variety of reservoir types (clastic sandstones, fractured carbonates, etc.) and field locations, it is hardly surprising that many different production scenarios are compared and implemented in

Table 6.1 Primary, secondary, and tertiary recovery methods.

Primary recovery	Secondary recovery	Tertiary recovery
Gas expansion	Water flooding	—
Solution gas drive	Gas compression/injection	—
Natural aquifer influx	Gas blow down	Water alternating gas (WAG)
Gas cap drive	—	—
Natural fractures: acid wash	Hydraulic fracturing	—
Directional drilling	Extended reach drilling (ERD)	—
—	Electric submersible pumps (ESP)	Thermal: steam flood
—	—	Chemical: polymer flood
—	—	Biological: microbial additive

a successful development project. In the simplest case, the production of hydrocarbons by natural depletion (primary recovery) involves single-phase flow by fluid expansion from the reservoir to the surface, through the wells: this might be gas expansion from solution or from a gas cap. This requires that pressure drawdown between the reservoir and the bottom of the well is sufficient to overcome viscous forces in the reservoir. It also demands that pressure drop over the wells to be larger than hydrostatic forces and frictions in the production tubing (Figure 6.3). When the pressure drop across the well is not large enough, artificial lifting methods (pumps, gas lift, electric submersible pumps [ESP], etc.) can be used to enhance the flow of the oil to the surface. These techniques are generally deemed mature, but progress is still required to increase the reliability of the lifting systems. This improvement could be achieved using intelligent systems that allow fully automated production.

Primary production in a fractured reservoir is often a function of well placement. The key to successful well placement is a thorough understanding of the fracture network. In many carbonate reservoirs the main interconnected fractures are separated by very low permeability blocks that, if penetrated by a well, will have a low production rate. However,

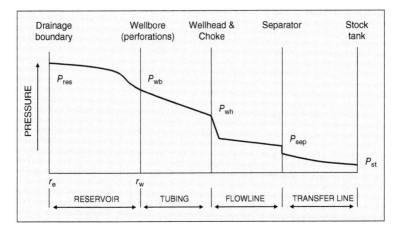

Figure 6.3 Pressure decline observed from the reservoir to the stock tank. This is the responsibility of a production or completion engineer to control and conserve the system energy. Too large a pressure drop can be detrimental to production.

connecting to a fracture corridor will result in near infinite permeability and high rates of production until the well waters out. The same might be demonstrated by a fractured basement or granite. Characterizing such reservoirs is huge challenge for the geologist and reservoir engineer.

The ability to drill extended reach wells to target bypassed or isolated accumulations has been part of the reservoir engineer's armory for the last 30 years. Previously, wells tended to be S-shaped from a central platform or drilling center giving an extended but limited radius of access, but with the advent of steerable motors complex well designs became feasible, reducing the number of drilling centers required. An example of the practical benefits of extended reach drilling (ERD) technology was developed at the Wytch Farm field in Southern England: a shallow Jurassic oil accumulation was produced from a number of onshore drilling centers, but when a large underlying Triassic oilfield was discovered extending into the middle of a picturesque harbor, tourist destination, and was overlooked by the most expensive real estate in England, an alternative development scenario was required. After rejecting an artificial island, approval was given for extended reach wells up to 10 km in length to be drilled from the existing surface facility, and for ESPs to actively produce oil (Figure 6.4).

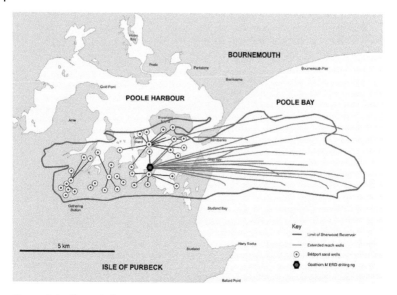

Figure 6.4 Extended reach drilling (ERD) at Wytch Farm to exploit the deeper Triassic Sherwood Sandstone reservoir. *Source:* Mike Simmons (2016); Neftex Exploration Insights; Wadiyat(2010).

6.2   Secondary Recovery

The first step in improving the oil recovery beyond natural depletion relies on injecting fluids that are initially not present in the reservoir for pressure maintenance. Although primary recovery may rely on the expansion of an aquifer or a gas cap that supports the pressure maintenance, in secondary recovery fluids are injected into the reservoir for this purpose. The choice of secondary and tertiary recovery methods depends on hydrocarbon mobility and reservoir heterogeneity: water flooding is most effective in homogenous sandstone filled with a light crude oil whereas heavy hydrocarbons require techniques such as steam flooding that improves oil mobility. Many fields are managed with a range of improved recovery methods over the lifetime; indeed, extending field lifetime is a key outcome of successful reservoir management.

There is a marked division between onshore and offshore fields; it is always cheaper to drill more wells either for production or injection onshore than offshore, and any retrofitting of equipment offshore also involves greater levels of investment. On the positive side, it is easier and cheaper to acquire a high-resolution 3D seismic survey offshore or to collect time-lapse 4D seismic data offshore to improve reservoir monitoring. The best way to address these many different techniques is through a series of case studies as we will see in Chapter 8. However, the starting point must be to look at the main methods first and where they are best applied.

6.2.1 Water Flooding

Water flooding is the most common secondary recovery displacement mechanism implemented in the industry: almost every oil field will sooner or later be water flooded if the right mobility conditions are observed. Economically, water flooding is often the most attractive way to increase ultimate recovery. Where a water flooding project is carefully designed and implemented, the ultimate recovery at the end of the process may be equivalent to a well-managed and efficient primary aquifer water drive.

Water flooding is relatively cheap, especially for offshore fields because of the ready availability of seawater, although care has to be taken to ensure that the injected water does not result in unwanted, adverse reactions in the reservoir. In some cases, injected brines may react with the naturally occurring connate water in the reservoir to form scale, while injecting very pure water rather than brine may result in clay swelling. Both of these may block the pores and reduce the permeability. The cost of drilling additional wells for injection is more than outweighed by the increased oil rates that result. Reinjection of gas (produced along with the oil) is used when there is no easy, economic way to export it for sale.

The main reason for water flooding an oil reservoir is to increase the oil production rate and, ultimately, the oil recovery. This is achieved by 'voidage replacement': injection of water to increase the reservoir pressure to its initial level and maintain it near that pressure, preferably above the bubble point. The injected water displaces oil from the pore spaces, but the efficiency of such displacement depends on many factors: hydrocarbon properties, microscopic oil displacement efficiency, rock/fluid properties, and reservoir heterogeneities. Ultimately, the recovery factor for water flooding is also determined by a number of external factors, including the architecture, number, and placement of water injection and production wells. Optimization of a waterflood project is dependent on these parameters to maximize the reservoir sweep and is an essential part of any profitable field development plan.

The displacement of oil is a function of the microscopic displacement efficiency due to capillary, viscous, and gravitational forces; the macroscopic sweep efficiency is the proportion of the connected volume of the reservoir contacted by the displacing fluid and the proportion of the reservoir connected to the producing wells. Within the field, due consideration must be given to any separate compartments created by faults or vertical barriers. These factors have been approximated in the equation:

$$\text{Recovery factor} = E_{PS} \times E_S \times E_D \times E_C$$

where E_{PS} are the microscopic forces, E_S is the sweep efficiency, E_D is the connected volume factor, and E_C represents the commercial effects on efficiency, that is, the constraints on field life due to the facilities or lease terms.

Oil properties are important to the technical and economic success of a water flood. The key oil properties are viscosity and density at reservoir conditions. In a porous medium, the mobility of a fluid is defined as its endpoint relative permeability divided by its viscosity; hence, a fluid with a low viscosity (≤ 1 cP) has a high mobility unless its relative permeability is very low. Similarly, a low-API crude oil ($\leq 20°$API) has a high viscosity and a very low mobility unless it is heated to high temperatures. Because the viscosity of water at reservoir temperatures is generally much lower than or, at best, equal to that of the reservoir oil, the water : oil viscosity ratio is generally much greater than 1 : 1. The water : oil mobility ratio is a key parameter in determining the efficiency of the water–oil displacement process, with the recovery efficiency increasing as the water : oil mobility ratio decreases.

In a large onshore field, there are a number of recognized water flood patterns that might be implemented: five-spot, seven-spot, and nine-spot or line drive are the commonest (Figure 6.5). Offshore, a system of producer-injector pairs is more common, or a simpler edge drive where the reservoir thickness is substantial, and heterogeneity is minimal. This represents a reduction in the number of wells required to improve

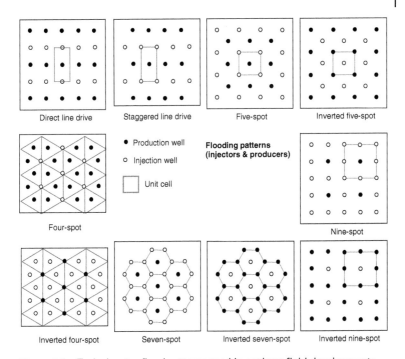

Figure 6.5 Typical water flood patterns used in onshore field developments.

recovery. Far better is to assume, as part of the development plan, that a water flood will be required and can therefore be planned; retrofitting of water handling facilities is expensive offshore.

However, there are numerous cases where primary operations are conducted in such a manner that water flooding is precluded as a secondary recovery process. For instance, if a gas cap is allowed or caused to shrink, and reservoir oil moves into the gas zone, much of that oil will be unrecoverable by water drive. Often, the majority of the oil that remains after primary operations is in portions of the reservoir rock not readily accessible to the encroaching water. The shrinkage and increase in viscosity of oil denuded of dissolved gas renders the oil less susceptible to displacement by water flooding. It should be recognized that not every old reservoir is a candidate for secondary recovery. Many secondary water flooding attempts have failed because a paucity of data or inept assessment failed to disclose the true nature of the prospect.

6.2.2 Gas Compression and Blowdown

As well as compressing gas to move it through a long pipeline, compression is used to improve recovery from gas wells as the reservoir pressure declines. A compressor is a mechanical device that is used to increase the pressure of a compressible fluid such as a gas or vapor by reducing the volume. The advantage of increasing the pressure of a fluid, but reducing its volume, helps to transport larger amounts of gas using a pipe of a specific size. Gas compression is also employed in gas storage and CO_2 sequestration projects.

Typically, gas compressors are sited onshore or on offshore platforms and used to boost production and transportation of gas to shore-based processing facilities. Increasingly, compressor units are located on the seabed where they are closer to the well head, thereby improving the efficiency of the unit. There are three types of engines used for compression: electric motor/centrifugal compressor, turbine/centrifugal compressor, and reciprocating compressor: the last two being powered by natural gas produced by the field, termed fuel gas.

Gas blowdown is a term used to describe the recovery of gas after the phase of oil recovery by a process of depressurization. The largest blowdown project was the Brent Field in the North Sea. After more than 25 years of oil and gas production it was decided to depressurize the field to release and recover solution gas from the bypassed oil in the reservoir, and to develop the secondary gas cap that had formed. The key element of depressurization reservoir management lies in systematic and regular monitoring and mapping of gas–oil and oil–water contacts, both areally and by layer, followed by reconciliation of the changes by material balance. Three of the four production platforms were converted for low-pressure gas production with a view to extending field life for 5–10 years. In 1998 all water injection ceased, and gas production commenced as planned.

6.2.3 Hydraulic Fracturing

Ignoring the environmental debate about 'fracking', the oil and gas industry has been using hydraulic fracturing to improve production for more than 70 years. The first commercial frac-job was carried out in 1949 by Halliburton using technology licensed to them by the Stanolind Oil &

Gas Corporation that had initially developed the concept but had limited success. Since then, fracking has been used as a well stimulation technique in conventional and, more recently, unconventional reservoirs. Essentially, hydraulic fracks are one part of a well completion strategy to maximize production from a reservoir. They can be used to induce fractures along the minimal principle stress direction or to enhance existing fractures through injection of inert proppant material such as sand or ceramic grains. A common application of fracking is to reduce 'skin' around a borehole due to formation damage during drilling and completion.

6.3 Tertiary Oil Recovery

Enhanced oil recovery (EOR) requires the application of an external source of energy or alternative materials to recover oils that would otherwise not be produced. There are three main classifications of techniques or methods: thermal methods, chemical approaches, and the use of miscible fluids (Table 6.2). The goal of EOR methods is to mobilize the residual hydrocarbon throughout the reservoir by increasing microscopic oil displacement and volumetric sweep efficiency: the same as primary and secondary recovery methods. Oil displacement efficiency can be increased by reducing oil viscosity through a thermal approach or reducing capillarity by introducing a chemical additive to the water flood. The addition of a polymer can improve the volumetric sweep efficiency. The cost associated with any enhanced or tertiary recovery project must be weighed against the potential outcome: how much additional oil can be recovered and at what cost. We will look at the cost in 'money of the day' in Chapter 7.

Table 6.2 Three main groups of tertiary recovery methods.

Thermal	Chemical	Miscible
Steam flooding	Polymer flooding	Water/gas flood
Steam stimulation	Surfactant/polymer	CO_2 flood
In situ combustion	Caustic flood	Nitrogen/flue gas

Where the mobility of a displacing fluid is greater than the fluid to be displaced then an unfavorable situation exists and the mobility ratio, based on the relationship between the two relative permeability quantities, is greater than unity. To improve the situation, it is necessary to lower the viscosity of the oil, increase the viscosity of the displacing phase, increase the permeability of oil, or reduce the relative permeability of the displacing phase. The other aspect of recovery is the capillarity of the system: the capillary number is a ratio of the viscous forces to the interfacial forces. Laboratory testing under reservoir conditions can determine the best conditions for a given rock type with a known porosity and permeability. Increasing the pressure gradient, reducing the viscosity of the fluid or interfacial tension can impact on the capillary effects in the reservoir. No single EOR process can be considered a panacea for improving recovery, but each in turn may have a positive impact. It is important that any preceding methods are considered before assessing the benefit of a new technique, and that a thorough screening of any process is carried out before application.

6.3.1 Steam Flooding

Thermal recovery methods are based on adding heat to the oil, mainly to decrease its viscosity. In this way, the mobility ratio between oil and the displacing fluids becomes more favorable. The most common thermal methods are steam flooding and steam cycling.

Steam flooding involves the continuous injection of 80% vaporized water to displace crude oil toward production wells. When steam enters the reservoir, it heats up the oil and reduces the viscosity; lighter components of the oil condense ahead of the steam front creating a piston to drive the oil to the producers (Figure 6.6). It is common practice to precede the steam drive by a cyclic stimulation of the producing wells, termed 'huff and puff'. A cyclic steam flood has three stages: first, injection during which a slug of steam is pumped into the reservoir; second, a soak stage, during which time the well is shut-in to allow the heat to be distributed in the reservoir around the well, and third, the 'thinned' oil is produced from the same well. Steam flooding is most effective in reservoirs greater than 20 m thick and comprising homogenous, high-permeability (> 200 mD) sandstone. It is most applicable where the oil gravity is less than 25 API and oil viscosity greater than 20 cP for the process to work. The efficiency of steam flooding is limited by gravity segregation of the fluids in thicker reservoirs. One way

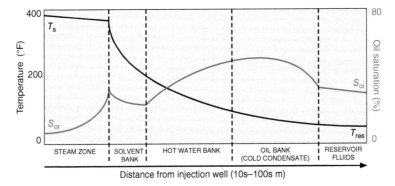

Figure 6.6 Subsurface regions of a typical steam flood operation showing the temperature profile and the oil saturation at each stage.

to mitigate this effect is to implement steam-assisted gravity drainage (SAGD): a pair of high-angled wells is drilled along a parallel trajectory with a vertical separation of around 5 m. Steam is injected into the reservoir through the upper well. As the steam increases oil mobility, gravity drives the oil into the lower producing well.

6.3.2 In Situ Combustion

This method, also known as a fire flood, involves starting a fire in the reservoir and injecting air to sustain the combustion of some of the oil. The heat generated produces hydrocarbon cracking, vaporization of the light hydrocarbons and formation water, while depositing the heavier 'coke'. As the front moves forward it pushes ahead a mixture of hot combustion gases, steam, and hot water that reduces the viscosity of the oil and mobilizes it toward production wells. This method is also most applicable where oil gravity is 10–25 °API and viscosity is less than 100 cP; some asphaltenes are required to produce the coke. This method is seldom a first-choice solution.

6.3.3 Polymer Flooding

A polymer flood aims to increase the viscosity of the displacing fluid (water) creating a mobility ratio greater than the oil, thus improving sweep efficiency by creating a smooth flood front without viscous fingering. The two most commonly used polymers are hydrolyzed polyacrylamide and xanthan. Ideally, the oil gravity should be greater than 25 API and viscosity preferably less than 100 cP. A polymer flood can be effective in heterogeneous reservoirs as the injected fluid flows through the higher permeability layers, reducing the flow rate, but equalizing the flow from the lower permeable layers.

Micellar polymer flooding reduces the interfacial and capillary forces between oil and water increasing oil mobility. The process requires a preflush of low-salinity water, a chemical solution comprising micellar or alkaline polymer, a mobility buffer, and a driving fluid, usually water that displaces the chemical and the oil bank to the production wells. Needless to say, the use of chemicals is expensive.

6.3.4 Caustic Flooding

Also known as alkaline flooding, this method involves the injection of sodium hydroxide, sodium silicate, or sodium carbonate. These chemicals react with organic compounds in certain hydrocarbons to create surfactants in situ: selecting the right type of hydrocarbon is essential. Polymers may be added to improve the sweep efficiency.

6.3.5 Miscible Flooding

Probably the commonest EOR methods used around the world, miscible flooding works by injecting a gas at high pressure into a reservoir to increase hydrocarbon mobility. Hydrocarbon miscible flooding involves injecting liquid petroleum gas (LPG), ethane-enriched natural gas, or lean gas to vaporize the light components from the crude oil being displaced. None of these methods is continuous; rather a period of injection is followed by the injection of water, hence the term water alternating gas (WAG) recovery.

Carbon dioxide flooding requires higher proportions of the gas to be injected to extract the lighter hydrocarbon components from the oil and if the pressure is great enough, forms a miscible flood. Nitrogen and flue gas flooding is a low-cost form of miscible recovery. Often, these gases are used to 'chase' the preceding hydrocarbon miscible or CO_2 flood.

All of these miscible flooding methods are limited to oils with gravity >30 API and viscosity <15 cP, with a high percentage of intermediate (C_5–C_{12}) hydrocarbons. Viscous fingering is a common problem in heterogeneous reservoirs resulting in poor vertical and horizontal sweep efficiency. The greatest disadvantage is the cost of the component products, especially LPG, and the poor recovery of the solvent gas. When non-hydrocarbon gases are used there is a cost involved in separating the saleable produced gas.

6.3.6 Other EOR Methods

Many laboratory techniques are being developed to improve hydrocarbon recovery, such is the value of the non-recovered oil; however, not all see the light of day in the field. Designer water or low-salinity water is a method developed by BP for use in previously water flooded fields. Trials carried out on the Endicott Field, North Slope Alaska showed a reduction in water cut after field had been flooded with a low-salinity water. The theory is that a wettability change is induced in the reservoir by multi-ion exchange (Seccombe et al. 2010).

Microbial enhanced oil recovery (MEOR), is the use of microorganisms to generate chemicals (surfactants, polymers, etc.) in the reservoir. MEOR relies either on injecting bacteria strands together with nutrients or on injecting nutrients to stimulate growth of bacteria naturally present in the formation (Bryant and Lockhart 2002; Awan et al. 2008). The specific surfactant- and polymer-generation processes depend strongly on the type of microorganism and rock and fluid properties. Extensive studies have been done to identify the most suitable microorganisms.

Finally, there is an extensive body of ongoing research devoted to developing new micro- and nanotechnologies in the oil industry. Fruition of such research could bring about considerable changes in the way oil exploration and production is done (Kong and Ohadi 2010). For instance, swarms of nanodevices transported by flood water could aid real-time mapping of reservoir fluids, resulting in unprecedented accuracy. Nanodevices are also being contemplated as carriers of chemicals (e.g. surface-active compounds) that can be delivered directly to the oil–water interface to modify the microscopic displacement pattern.

6.4   Summary

Many great ideas have been tried on many, many fields to try and increase recovery and extend field life. These projects have kept engineers in work performing trials and studies but in the end it all comes down to economics. One of the latest projects that is truly innovative is GlassPoint Solar in Oman. This is a large-scale solar powered project designed to save gas needed to fuel the country's economic growth; however, at the current gas price (2020) this project is on hold.

7

Development Economics

The oil and gas industry is one in which operators take long-term decisions on investment, but expect a relatively quick return because they need the money to invest in the next project: companies need to replace their hydrocarbon production to continue to do business. So, to help in the decision-making process there are three simple questions to answer:

- What will it cost?
- What is it worth?
- Will it return sufficient profit quickly enough?
  Any viable business needs to make a profit to continue trading:

$$\text{Profit} = \text{Revenue} - \text{Cost}$$

In a failing business, if revenue is less than cost then a loss is the outcome. There are three main ways to describe profit: net cash flow, financial net income, and a tax model. These simple truths are complicated by taxes, depreciation, expenses, and time. What is the value of the project at the time the investment is made and what is the rate of return on that investment: also termed cash flow? A period of time, usually either a month or a year in the oil business, is defined to measure the financial success of a project.

Before a sensible economic valuation of a project can be made it is essential that the following data is known or can be estimated:

- Production rate against time.
- Oil and gas price.

- Capital expenditure (CAPEX) and operating costs (OPEX).
- Royalty payable or production sharing costs.
- Discount rates and inflation.
- National or local taxes (petroleum revenue tax).

For some of the criteria, estimates or ranges must be given but at the time of project sanction to incorporate the time value of money.

7.1 Key Economic Criteria

As part of the development project an economic objective based on the company's economic criteria must be set. This usually requires information on the internal rate of return (IRR), any hurdle rates based on interest charges, and also a company's attitude to risk.

7.1.1 Net Cash Flow

To evaluate an investment opportunity, be it an infill well, compression project, or a full-blown field development, the after-tax profit (or loss) is expressed as the present value of cash flow or net cash flow and is measured over the life of the project. The net cash flow model is the most commonly used in the industry to forecast profitability of a project until such time as it is deemed uneconomic: this is not a lump sum but the profits over the life of the project. Although project evaluations are done on a monthly basis, they will normally only be reported annually to smooth out any planned or unplanned downtime. The present value of a cash flow stream after tax is expressed as:

$$\text{Net cash flow} = \left[\begin{pmatrix} \text{NRI} \times \text{Production} \times \text{Price} \end{pmatrix} - \\ \begin{pmatrix} \text{Wellhead taxes} - \text{OPEX} - \text{Taxes} - \text{Investment} \end{pmatrix} \right]$$

An operator will usually pay all of the cost of a project but only get a proportion of the revenue from production; this is known as net revenue interest (NRI) and is a result of a royalty paid to an owner of the property or the local or national government as a petroleum revenue tax.

Net cash flow is normally discounted against the value of money of the day so that different projects can be compared with other investment opportunities. Expressing the value of a project in terms of the present-day value indicates that it is discounted to reflect the time value of the money.

7.1.2 Time Value of Money

The time value of money is important in the oil and gas industry because we need to know how to place a value on our investment received as a result of production in the future. There are two ways to look at the future value of money: one by estimating the interest going forward and the other looking at the effect of inflation. In other words, what is my dollar/pound/euro going to be worth if invested at a fixed rate of return or how much will I receive in the future at a given inflation rate (Table 7.1).

$$\text{Future value} = \text{Present value} \times (1+i)^t$$

$$\text{Present value} = \text{Future value} / (1+i)^t$$

Combining the concepts of net cash flow and time value of money leads to a ubiquitous cash flow time diagram: profit is measured as net cash flow in or out (expense) annually, except for time zero when the first investment is made (Figure 7.1). By using the present value approach above it is possible to calculate the net present value (NPV) of a project: NPV is equivalent to the project cash flows at the assumed discount rate.

7.1.3 Payout Time

This is the time required to recoup the initial investment. This is the time when the undiscounted or discounted cash flow is equal to zero.

Table 7.1 The time value of money expressed in terms of interest and inflation to estimate net present values of a project.

Time (t)	$t = 0$	$t = 1$	$t = 2$	$t = 3$	$t = 5$	$t = 10$
Interest (10%)	$ = 1	$1.10	$1.21	$1.33	$1.62	$2.59
Inflation (10%)	$ = 1	$0.909	$0.826	$0.751	$0.62	$0.385

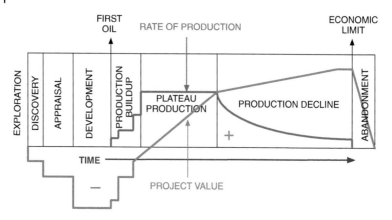

Figure 7.1   Oilfield life cycle showing the production profile from buildup to abandonment for primary recovery with a comparison of the project value, which is negative until production has started.

$$\text{Cashflow} = \text{Revenue} - \text{Capital investment} - \text{Operating expenses} = 0$$

The shorter the payout time the more attractive a project becomes to the investors: in the oil and gas business this is commonly two to five years. Although a simple way to look at a project, it does not focus on the longer-term viability of a development and should not be the only measure of success. *Discounted cashflow* is used to account for the time value of money by converting the future worth of money to the NPV: this is not required if the undiscounted cash flow is used. If revenues are received midway through the annual accounting period, the discount factor is equivalent to:

$$\text{Discount factor} = \frac{1}{\left(1 - i\right)^{t - 0.5}}$$

where t is the time and i is the discount rate as a fraction.

7.1.4   Net Present Value

Net present value (NPV) is equivalent to the future monthly or annual cash flows at the assumed discount or hurdle rate. It is the most commonly-used parameter to express the value of a *successful* project, and it measures the cumulative cash worth of the venture above the corporate discount rate. After the discounting method has been selected, there remains the question of the actual rate. The average investment opportunity rate is the interest rate that represents the return on the future investment opportunities available to the company: this is why it is most often used as a screening tool. When the cash flow of the operation is discounted at the given percentage rate, a positive or negative value is calculated. For project screening all positive outcomes are acceptable; if the projects perform as expected they will return more to the company than the average company project will return. For NPV to be useful, the present value date and the discount rate must be specified. The application requires choosing the project or investment with the highest positive value of NPV in preference to the lesser. NPV is normally based on the mean reserves case.

7.1.5 Internal Rate of Return

The IRR is defined as the interest rate that causes NPV calculations to return a zero value. IRR can also be used to screen projects: if the IRR is greater than the average investment opportunity rate, the project passes the screen (Figure 7.2). IRR must only be used on comparable projects, and not those that might be mutually exclusive. It is possible that more than one interest rate can cause the NPV to be zero: this is termed multiple rates of return and usually occurs in acceleration projects. An acceleration project is one where an investment is made to speed up the payout on a development; these generally result in an overall loss.

7.1.6 Maximum Negative Cash Flow

Maximum negative cash flow (MNCF) represents the greatest cumulative 'out-of-pocket' expense associated with a project; the maximum financial exposure in any project. MNCF is useful in budgeting, planning, and for comparing a portfolio of potential projects. Smaller, cash-constrained companies use the criteria to judge their financial exposure to a project. The term used is 'undiscounted' and is calculated from the cash flow model by expressing the net of the investments and cost against early revenues and can be related to the time to payout when a project

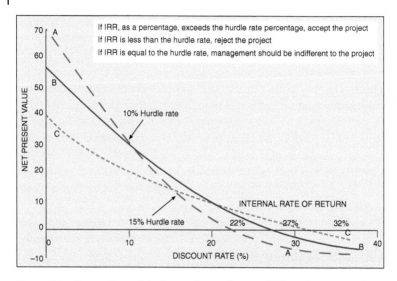

Figure 7.2 Comparison of the internal rate of return (IRR) on investment for different hurdle rates. Only accept projects that show a positive rate of return.

goes from negative to positive earnings. This criterion does not address the chance of success but can give some comfort to the more risk averse investor.

7.1.7 Profit-to-Investment Ratio

Profit-to-investment ratio (P/I) may be undiscounted or discounted. If undiscounted numbers are used, the result is an undiscounted P/I; if discounted numbers are used, the result is a discounted P/I. Undiscounted P/I is the total cash flow, without capital investment, divided by the total investment; it reflects the total profitability but does not recognize the time value of money. Neither does it express the scale of a project or any aspect of risk.

Discounted P/I is calculated by dividing the sum of either the net operating income or the net cash flow from a project by the sum of the investments. If net operating income is used in the numerator, a value of 1.0 is a breakeven value where the investment is just recovered. If net cash flow is used in the numerator, a value of 0.0 is a breakeven value. Either definition is appropriate for the numerator, as long as it is clearly stated which definition has been used.

Discounted P/I at the company average investment opportunity rate is a powerful selection and ranking tool. As a selection tool, all projects with a value greater than 1.0 (or 0.0) would be selected. In the presence of limited capital, the projects are ranked in decreasing order of discounted P/I and selected until the capital available for investment is exhausted. This very simple tool results in the portfolio of projects that causes the treasury to grow at the fastest rate, if the projects perform as expected. This approach maximizes expected value but ignores risk. In fact, funding projects with the highest discounted P/I will tend to produce a high-risk portfolio.

7.1.8 Investment Efficiency

Many of these economic parameters can be used to measure the efficiency of an investment, but it is essential that when comparing projects the same measure is used in all cases. Two of these methods are investment efficiency and discounted P/I as discussed in Section 7.1.7. The first is the ratio of the cumulative NPV against the present value of the MNCF, and the second is cumulative NPV against the present value of all investments: both are therefore discounted properties.

7.1.9 Building a Cash Flow Model

Using a spreadsheet, it is relatively straightforward to build a cash flow model, as long as you have all the necessary input data. The cash flow model must include all the upfront costs including investment in exploration: dry hole cost, recoverable reserves, number of wells required, and a typical production schedule to abandonment. Add to this the development costs including drilling cost, operating cost per well, well head prices, taxes, transportation (pipeline fees), all taxes, NRI, company discount rate, and any incremental income tax. Not all of these will apply to every development, but there may be other charges such as in a production sharing agreement (PSA).

7.2 Risk and Uncertainty

Individuals are all introduced to the concept of risk early in life and most people are risk averse: you don't try to cross a busy road unless there is a pedestrian crossing, or do you? I once saw a young lady in a hurry try to beat the lights when crossing a road; she would have been fine had she not tripped and fallen. Fortunately, the cars were slow to start and were able to stop in time, but she had taken a risk that always makes me think twice. How much money are you prepared to risk on the toss of a coin? I was once told never to gamble more than I could afford to lose. However, oil companies and their investors exist to take risk, but people are very inconsistent in evaluating risk; consider the exploration geologist and the development engineer: they will consider risk in very different ways. When comparing a single exploration well against a project to increase oil recovery the difference is apparent: a low-cost but high-return against a high-cost, low-return opportunity. And that's before the bean counters (accountants) get involved. What is required is a common and consistent way to compare the investment opportunities; however, we all have an internal bias founded on our own experience, knowledge, and background.

Risk and uncertainty are two very different aspects of the same decision-making process. Risk is always present in any decision we make as there is always a possibility of a negative outcome. Risk is a measure of a discrete outcome: the chance of success against the chance of failure. Uncertainty is a way to try to cope with that risk and measure the range of positive and negative outcomes of the decision; it describes a range of equiprobable outcomes. The industry has developed a number of different analytical methods to measure risk and uncertainty (Table 7.2).

Within the reservoir management context, the primary uncertainty is in the original hydrocarbon in-place estimation, with the risk being that there is less gas or oil present, or that the ultimate recovery is compromised by the chosen development plan. Uncertainty can be reduced through time and production, but decisions are always made with an incomplete understanding of the subsurface. Equally, there is the issue of remaining reserves uncertainty, which changes with time, and whether the value of remaining reserves should be increased incrementally or through accelerated production (Figure 7.3).

There are different types of uncertainty: randomness, fuzziness, and incompleteness. Randomness is present when there is the lack of a specific pattern: a uniqueness in every outcome. Fuzziness implies an imprecision of a defined variable and incompleteness is a lack of information. Uncertainty can also be expressed as statistical (aleatoric) or systematic (epistemic). Statistical uncertainty reflects an inherent randomness in the data due to natural variation: over time all outcomes

Table 7.2 Different biases affecting risk decisions.

Framing effects	Decision-makers will take a greater gamble to avoid a loss than to make an equal gain
Existence of a prior account	Decision-makers are more likely to take a risk at the start of a project than later in the project's life
Maintaining a consistent frame of reference	Decision-makers are more likely to invest during a run of good fortune than when having a bad run of luck
Probability of success	A venture which is perceived to have a high probability of success is preferred to one which is riskier even though the expected value of the latter is greater
Wrong action vs. inaction	Managers avoid criticism by not making a decision rather than taking action that could result in the same loss
Number of people making a decision	Groups are more prone to risks than individuals
Workload and venture size	Large volume ventures are preferred over smaller ones, especially when decision-makers are busy
Personal familiarity	Decisions made in the 'comfort zone' are always easier to make!

will be experienced. Systematic uncertainty results from a lack of knowledge or inadequate understanding of the data; however, this can be reduced by gathering additional data.

Most treatises on petroleum economics discuss risk and uncertainty in terms of exploration decisions; should you run more seismic or just drill a well? This is to better define the drill or drop decision. This is seldom the case in field development. The commonest decision-making tool is to generate a risk register in which all the uncertainties are quantified and solutions to mitigate that uncertainty recorded. The final decision will generally be one of cash flow and NPV. The foundation of risk analysis in this context is expected monetary value (EMV): this is a means of consistent decision-making rather than just profitability of a project.

EMV is a method of combining quantified estimates of each element of an investment opportunity and to assign a probability for each outcome. This gives a risk-adjusted decision-making solution. The

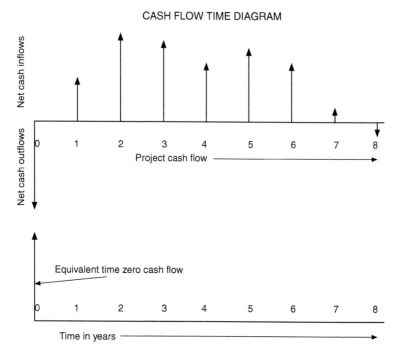

Figure 7.3 Cash flow time diagram showing the net cash outflow at the start of a project compared with the years of net cash inflows until abandonment.

parameter being analyzed is computed as the sum of the mathematical product of the probability of each outcome for all the possible outcomes. The method requires two or more possible outcomes for each alternative solution, but the outcomes identified must include all the possible outcomes for the option being evaluated. Each outcome must have the possibility of occurring, but none can be guaranteed. The assigned probabilities must be proportional to the possibility of occurrence, and the sum of all probabilities must equal one.

$$\text{Expected value} = \left(R_1 \times P_1\right) + \left(R_2 \times P_2\right) + \left(R_3 \times P_3\right) + \dots \text{where } P_1 + P_2 + P_3 = 1$$

The results (R) can either be positive or negative and can occur in various combinations depending on the project and the number of possible outcomes. The EMV can be seen as the mean outcome anticipated from a large number of similar projects. In this way EMV can be used to value a number of investment opportunities in a company's portfolio.

7.2.1 Probability

Probability is the mathematical study of uncertainty; the likelihood of an event occurring. Probability must be between 0 and 1, with higher values indicating an increased possibility. The sum of all probabilities of occurrence must equal unity (1), and outcomes that are mutually exclusive cannot both occur. It is common in many situations to assign a numerical probability to an event when there is insufficient statistical data to determine an outcome. With experience, it may be possible to estimate a range of outcomes using some numerical value: a project may cost between $270–320MM, based on an estimate of $290MM, +/− 10%. Human beings are notoriously bad at estimating because of inbuilt bias; generally, we are either too optimistic or too conservative depending on the situation, especially if there is a fear of criticism or a censure for a bad choice. There is a special case when the estimate is driven by self-interest, such as a bonus or promotion: this will always result in an overestimation.

Using probability theory to define risk requires some knowledge of how that probability is distributed and also on what it is based (Figure 7.4).

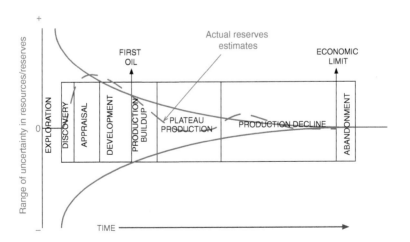

Figure 7.4 Range of uncertainty in reserves of time of compared with the stages of an oil field life cycle. You will only know the ultimate recovery from a field when it is abandoned.

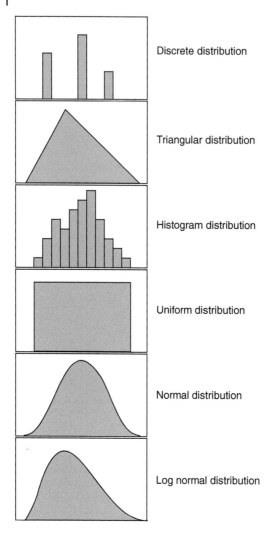

Discrete distribution

Triangular distribution

Histogram distribution

Uniform distribution

Normal distribution

Log normal distribution

Figure 7.5 Different statistical distributions that may be used in stochastic calculations of reserves for the different input variables.

- Discrete: single values drawn from a number of similar events.
- Triangular: where the discrete values define a triangle.
- Histogram: where there are sufficient events to be binned into classes.
- Uniform: where there is a single value for a range of outcomes.

- Normal: is defined by a mean and standard deviation of many representative samples.
- Lognormal: a positively or negatively skewed distribution.

If a proper data partitioning analysis is undertaken, uncertainty is not normally distributed but skewed: looking at a porosity, for instance, all samples analyzed will probably show a normal distribution but by partitioning the data into facies or rock types will result in different distribution. Lognormal distribution arises when multiplying parameters such as porosity feet, field size, or production rates (Figure 7.5).

7.3 Summary

Petroleum economics is a subject worthy of much study by accountants and asset managers, but the bottom line will always be 'the bottom line': how much is it worth to my company today and what will it cost to make the project fly? The accountants will always want to spend the minimum and if the asset manager doesn't accept the need for additional information you could end up with a project that costs a lot more than anticipated: see Chapter 8.

8

Tales of the Unexpected

There are around 2000 producing oil, condensate, or gas fields scattered around the world, mostly in some of the more inhospitable places known to humans from the deserts of the Arabian Peninsula to the North Sea and Alaska's North Slope. Each field has a history, many more than 100 years old, and it is the aim of this chapter of the book to look at some of the success stories and unexpected events that have been reported in the academic and commercial journals. I have been fortunate to have worked in some capacity or other on all of these fields and so have reasonable knowledge of their history and stories!

8.1   Laggan and Tormore, Flett Basin, West of Shetland, UKCS

The story of Laggan and Tormore is a long and complicated one with many a twist and turn. The Laggan Field was discovered in 1986 by Shell with well 206/1-2; the first well on the block was a failure. The well was tested, achieving a rate of 25 MMscf/d dry gas. Ten years later, Total drilled appraisal well 206/1-3 and established a GWC that was later also identified on seismic as a marked amplitude change. The impact of modern seismic acquisition and processing influenced much of the story of these two fields. Further appraisal drilling took place in 2004 that established the updip extent of the field: wells 206/1a-4A and 206/1a-4Z. Based on a similar

Reservoir Management: A Practical Guide, First Edition. Steve Cannon.
© 2021 John Wiley & Sons Ltd. Published 2021 by John Wiley & Sons Ltd.

seismic response in the adjoining block a further well was drilled in 2007, 205/5a-1 that discovered the Tormore Field; a gas accumulation with moderate liquid content (Figure 8.1). The well was tested and produced 28.3 MMscf/d from one of three sandstone reservoirs. The sands penetrated in both fields are Paleocene Vaila Formation, comprising stacked turbidite bodies with excellent reservoir properties: porosity 19–24% and permeability 20–200 mD. The quality of the reservoir is considered exceptional because of the depth of burial (3000 m). The sand grains are coated with chlorite that seems to have preserved the pore space during burial.

Based on an in-place estimate of ~1 TCF (230 MMBOE) in the two fields, a development plan was approved in 2010 with an initial budget of ~£2.5 billion. The plan included two 6-slot manifold templates, a semiautonomous subsea production system, a 145 km dual pipeline and control

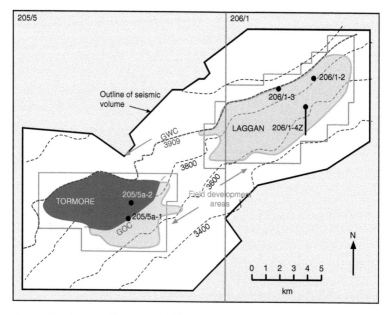

Figure 8.1 Laggan–Tormore Field location map with predevelopment well locations, field development areas, and gas–water contact. Well 205/5a-2 is the appraisal development well that encountered an oil leg unexpectedly. *Source:* Redrawn from Noel, P. and Taylor, N. (2016) Beyond Laggan-Tormore: maximising economic recovery from gas infrastructure West of Shetland. In: Petroleum Geology of NW Europe: 59 Years of Learning – Proceedings of the 8th Petroleum Geology Conference, 455-464. Copyright 2018 Petroleum Geology Conferences Ltd. Published by the Geological Society, London.

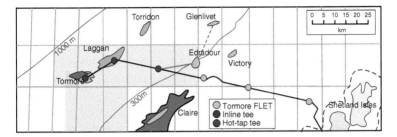

Figure 8.2   The west of Shetland gas export route from Laggan–Tormore to Sullom Voe gas terminal showing the different tie-in points. *Source:* Redrawn from Noel, P. and Taylor, N. (2016) Beyond Laggan-Tormore: maximising economic recovery from gas infrastructure West of Shetland. In: Petroleum Geology of NW Europe: 59 Years of Learning – Proceedings of the 8th Petroleum Geology Conference, 455-464. Copyright 2018 Petroleum Geology Conferences Ltd. Published by the Geological Society, London.: modified and redrawn).

system, and an £800 mm gas processing plant on the Shetland Islands (Figure 8.2). The field is in 600 m of water and in some of the harshest weather conditions, which is why a subsea completion, the deepest in the UK, was chosen as the best solution. The two 18-in. multiphase pipelines also represent the longest deepwater transmission system and presents many flow assurance issues not least the slope of the seabed to the continental shelf. The two pig-catchers required for "pigging" the lines could be seen from space! The new gas processing plant at Sullom Voe is designed to handle 500 MMscf/d, with condensate being stripped and sent to the existing oil terminal operated by BP. The gas is then exported to the mainland via a new pipeline, SIRGE, and the existing FUKA system to St. Fergus. The final cost of development exceeded £3.5MM but will provide a long-term export route for gas from the west of Shetland. Key to this is a number of in-line tees in the pipeline to which any further discoveries could be routed once the initial period of peak gas production ended and there was ullage in the system.

Development drilling began on Tormore in 2012 with a well downdip of the discovery well; the well found a 35 m oil column, not gas! Four gas producers were subsequently drilled on Laggan and tied into the production system. First gas to shore was achieved in February 2016 fulfilling the initial 500 MMscf/d plateau production. The challenge would be to maintain the plateau for as long as possible, given the surprising results at Tormore. This is where the in-line tees come in to play; the operator had a number of leads and discoveries in the area that could be tied-in if economical. The first three exploration wells, drilled largely on the basis of seismic amplitude anomalies, revealed varying results, none of which turned out to be economical. The operator then started to engage with other companies with proven discoveries and was able to do an equity deal with DONG Energy regarding the Glenlivet Field. Two producers were drilled and tied back to the Laggan–Tormore pipeline fulfilling the production requirements for the whole project. Total and its partners continue to explore for additional opportunities to maximize the 20-year life of the Shetland Gas Plant (Noel and Taylor 2018).

8.2 Dación Field, Maturín Basin, Venezuela

The Oficina trend in northeast Venezuela is one of the world's major hydrocarbon basins and the Dación Field one of the jewels: since 1940, it has produced more than 300 MMbbls of oil and has estimated remaining reserves of 1.9 Bbbls. In 1998, as part of the third rehabilitation round, the state-owned PdVSA (Petroleos de Venezuela) awarded a long-term development contract to a joint venture of LASMO and Schlumberger: the former would manage the implementation of the development plan, and Schlumberger would handle the drilling and production operations. LASMO won the right to develop the field by paying $453 mm, around $100 mm more than the second highest bid. LASMO believed that there were substantial exploration opportunities in the structurally and stratigraphically complex reservoir, while Schlumberger would be rewarded based on increasing production above an agreed baseline. When the alliance took over the field production was at 13000 bbl/d and over the following three years increased to 42000 bbl/d.

A 3D seismic campaign over the entire contract area was completed in 1999, the results of which provided a block-wide view of the subsurface for the first time, revealing multiple prospects in a wide variety of new and existing play settings. A planned 17 well exploration drilling program was designed to test the highlighted prospectivity (Frorup et al. 2002). One of these wells, Tortola-1, was drilled to a total depth of 2082 m and encountered 21 m of oil and gas bearing pay in the Middle–Lower Miocene Oficina formation. Four oil zones were tested, achieving a cumulative flowrate of 3600 bpd of 22–25 °API oil. The Tortola discovery is located in the southeastern area of the Dación block within 3.5 km of existing production facilities.

Operational performance led to drilling times being reduced substantially from 20 days to less than 10 days, with costs falling in-line, such that the cost to drill and complete a well was just over $1 mm in 2000. Other technological breakthroughs reported include multizone gravel pack completions, centrifugal water separation, and multiphase metering. These results demonstrated the benefits of an integrated project management approach to field development.

In 2006, the Venezuelan government decided to take back control of several fields operated by international companies, Dación being one of them. The field was then owned by ENI, who acquired LASMO in 2001. The foreign energy companies were required to convert their operating contracts into joint ventures with PdVSA. In 2008, the company came to a financial rapprochement with the Venezuelan authorities, presumably Schlumberger are still involved in field management.

8.3 As-Sarah Field, East Sirt Basin, Libya

As-Sarah Field is located onshore in the Maragh Trough of the East Sirt Basin, Libya. It was discovered in 1989 and came onstream in 1990. In-place volumes are estimated as greater than one billion barrels of oil and recoverable reserves of 500 MMBO, giving a calculated recovery factor of about 50%. A light oil pool is trapped in a west-northwesterly-trending, northern-dipping fault block (Figure 8.3). Extensional faulting began in the Triassic and may have extended into the Late Cretaceous when maximum subsidence occurred. The reservoir interval consists of stacked fining-upwards braided fluvial sequences and becomes finer-grained and more siltstone-prone, with increasing soil pedification, toward the top, reflecting a less active fluvial system through time. Permeability is typically 80–500 mD (average of 100 mD) and reservoir quality is controlled by depositional facies, with high-energy channel sands forming the best reservoirs.

The reservoir is internally well-connected, with the only significant barrier to flow consisting of a tar mat at the oil–water contact (OWC). This prevented pressure support from the underlying aquifer during the first four years of production but after this broke down when the pressure differential between reservoir and aquifer disappeared, the planned water injection scheme was unnecessary. The field has been developed by three rows of wells parallel to the axis of the structure. The first row is

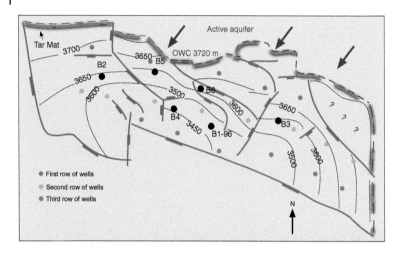

Figure 8.3 Structure map of the As-Sarah oilfield, Libya, showing the active aquifer and the first, second, and third sequential production wells. *Source:* Redrawn from Bremeier et al (2005).

positioned lowest on the north flank, and its wells will water out first as the aquifer encroaches from the northwest. As this occurs, they are shut-in and production takes place increasingly from the second and third row wells located updip. This scheme has facilitated a plateau production of ~100 000 BOPD, which was predicted to end shortly after 2007. Artificial gas lift was implemented in 2003 to optimize recovery of wells with high water-cuts, located downflank in the north of the field. The reservoir has a very low initial water saturation of 1–6% and is oil-wet, potentially making it favorable in the future to enhanced oil recovery (EOR) techniques (Bremeier et al. 2005).

Conditions in war-torn Libya have prevented further development of the field, which was shut-in for a number of years.

8.4 Ceiba Field, Rio Muni Basin, Equatorial Guinea

The Ceiba Field is in the Rio Muni Basin, offshore Equatorial Guinea, in 670–800 m of water (Figure 8.4). It was discovered and put onstream in 1999 and 2000, respectively. It has reserves of 245 MMBO light oil trapped in a northeastern-trending anticline by dip closure, stratigraphic

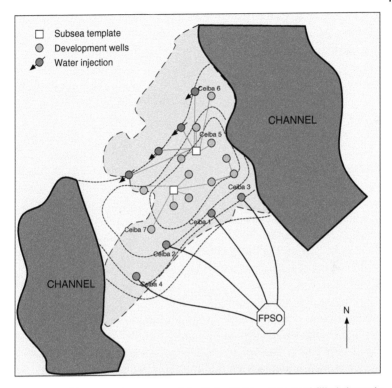

Figure 8.4   Top reservoir structure of the Ceiba Field and the mud-filled channels that form the lateral trapping mechanism. Also identified are the locations of the main in-field subsea infrastructure elements. *Source:* Redrawn from Dailly, P., Lowry, P., Goh, K. and Monson, G. (2002) Exploration and development of Ceoba Field, Rio Mini Basin, Southern Equatorial Guinea. The Leading Edge, November 2002 1140-1147. Society of Exploration Geophysicists: modified and redrawn.

pinch-out, and unconformity truncation. Two Tertiary mud-filled canyons erode into the northeastern and southwestern margins of the field, providing additional trapping elements. Campanian sandstone reservoirs comprise: (i) thick amalgamated, turbidite-channel sandstones; (ii) finely laminated, overbank and interchannel sandstones and siltstones; and (iii) deeper (Campanian) sandstones with limited distribution.

The field has four OWCs that reflect a labyrinthine reservoir architecture, but several of the uppermost and thickest channel-fills amalgamate into what is known as the Main Pool, which locally has a tank-like reservoir architecture and provides most of the field's production. Main Pool porosity and permeability average 26% and 500 mD, respectively. A fast-track development program brought the Ceiba Field onstream 14 months after discovery. The first phase of the development from 1999 to 2001 consisted of two exploration/appraisal wells and two development wells to exploit the southeastern flank of the field where the Main Pool comprises channel sands that are in communication. The second phase, in 2002, added six new production wells and four peripheral water injectors, focusing on the western flank of the field.

Wells benefit from hydraulic fracture treatment. The Ceiba-5 well, which is located on the northern flank of the field, flowed at an initial rate of 29 912 BOPD and 15.18 MMCFGPD, the highest recorded in the field (Claiborne et al. 2002). This high rate benefited from a frac-pack treatment across two distinct screenout intervals. Fracture geometries were complex owing to lengthy perforation intervals in wells that were deviated by as much 60° (Cipolla et al. 2005). The fracture treatments had the added benefit of reducing wellbore skin from as much as 10–15 in the earlier wells to 0.8–1.3 in the Ceiba-5 well. The Ceiba-7 well, which was completed in the laminated sand facies, was given a hydraulic fracture treatment, combined with an open hole gravel pack completion, leading to higher productivity.

Oil is piped 7 km to a floating production, storage, and offloading vessel (FPSO) that is moored in 90 m of water. Facilities were designed for 135 000 BOPD production, but oil production was very variable during 2001, ranging from a high of 50 000 BOPD in April to 20 000 BOPD in December. After the changeover of the FPSO from the Sendje Berge to the Sendje Ceiba in January 2002, production jumped to 60 000 BOPD. Production, which was predicted to reach 90 000 BOPD by Q1 2002, then slumped to around 30 000 BOPD by late 2002. This occurred because of a 16-month delay in starting water injection, the high initial offtake rates under primary production, and the unexpectedly complex reservoir geometries. Production in 2005 averaged 40 000 BOPD compared to expectations of 135 000 BOPD, while 120 000 BWPD was injected into reservoirs. In 2006, production from the Ceiba Field was ~33 000 BOPD (Amerada Hess 2002). In 2017, the field was sold to Kosmos Energy with an expectation of increased production rates that reached 44 000 BOPD, but it proved hard to maintain this increase, probably due to a more complex reservoir architecture than anticipated.

8.5 Glenn Pool Field, Cherokee Basin, Oklahoma, USA

The Glenn Pool Field is located in the Cherokee Basin, Northeast Oklahoma, USA. The field was discovered and came onstream in 1905; peak production of approximately 117000 BOPD was achieved in 1907. In-place and recoverable reserves are 1570 MMBO and 400 MMBO, respectively, giving a 25% recovery factor; average reserves are estimated at 10000 BO/ac. A single, light oil pool and several small gas pools are contained in a stratigraphic trap created by updip pinch-out of fluvial sands (Figure 8.5).

The main reservoir is the Middle Pennsylvanian Bartlesville Formation of the Cherokee Group, locally known as the Glenn (Sandstone) Formation. Traditionally divided into lower, middle, and upper units/ members, sequence stratigraphic studies since the mid-1990s have developed a more refined reservoir scheme. The reservoirs consist of very fine- to medium-grained sub-litharenites and litharenites deposited in a wide (~40–50 mi), S-shaped incised fluvial valley. Initial production was by natural depletion, followed by pumping; however, gas and water injection

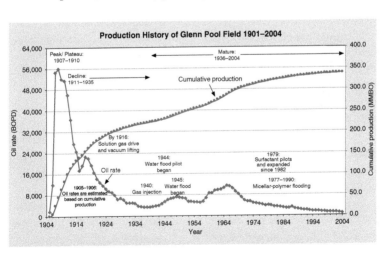

Figure 8.5 Glenn Pool production profile showing both annual and cumulative production with the timing and impact on production of secondary and tertiary recovery methods over 100 years of field life. *Source:* Redrawn from Kuykendall and Matson (1992).

were introduced in 1940 and 1955, respectively, to improve pressure management, leading to secondary production peaks in the late 1940s and mid-1960s. A micellar-polymer injection project was initiated in 1980, and in the mid-1990s localized redrilling and selective perforation, coupled with increased water injection, led to higher production rates. By 2004, only 21.5% of the in-place reserves had been produced and ultimate low recovery is attributed mainly to significant reservoir heterogeneity; production in 2004 is around 1000 BOPD. Future production is anticipated to come from complex meander and crevasse-splay sand facies.

The field was rapidly developed after its discovery in 1905 and annual production peaked in 1908 at 20.5 MMBO (~56 000 BOPD), with initial well production rates of 75–4000 BOPD (Kelkar and Richmond 1996). Wells were initially randomly spaced and later drilled on a 10 ac spacing. Storage facilities were limited, and oil was routinely stored in newly constructed lakes. The oil was initially produced by solution gas drive, but rapid pressure depletion of the reservoir meant that by 1916 most wells required pumping. By 1939, production had declined to <1000 BOPD and secondary recovery by recycled gas injection was introduced in 1940 on a 160 ac pilot area. Production rose from 61 BOPD to 300 BOPD and the scheme was extended to include 2000 ac in the southern part of the field by 1943. Infill drilling and gas injection succeeded in increasing field production throughout the 1940s. Several single-well water injection tests were conducted during the 1940s, but the first successful trial took place in 1953 in the north of the field and led to significant production improvements within two years of its implementation. In 1955, the first multi-pattern water injection scheme was initiated, and its success led to its field-wide implementation, resulting in substantial improvements in recoveries from 5000 BO/ac to over 8000 BO/ac. A secondary production peak of nearly 5 MMBO per year (~13 700 BOPD) was reached in 1965, but production declined rapidly to <4000 BOPD in the mid-1980s (Kuykendall and Matson 1992) and was 1000 BOPD by 2004. By 1999, the Glenn Pool Field had produced 336 MMBO, or 21% of the in-place reserves (Kerr et al. 1999a). After the implementation of each secondary technique the reservoir has responded with an increase in production, indicating a high degree of heterogeneity and suggesting that the field will continue to deliver with further reservoir management (Ahuja et al. 1994).

EOR pilots, principally involving surfactant flooding, have been implemented in a number of units in the Glenn Pool Field. Micellar-polymer injection tests were carried out in the 160 ac William Berryhill unit in the south-central part of the field, beginning in 1977. By 1990, this EOR project had recovered 1.2 MMBO, compared to the 1.1 MMBO recovered by water injection from 1955 to 1977. The production rate in the project area increased from 50 to 60 BOPD under water injection to 1200–1500 BOPD under micellar-polymer injection (Kuykendall and Matson 1992).

The Self Unit (Figure 8.6), located in the south of the field, has also been subjected to gas and water injection (beginning 1945 and 1954, respectively) and in 1978 underwent a redrill program that selectively

(a)

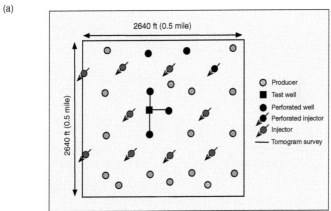

Self unit well location map (Kelkar and Richmond, 1996)

(b)

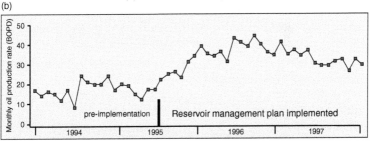

Increase in production in the Self unit after the implementation of the reservoir management plan (Kerr et al. 1999).

Figure 8.6 (a) Self Unit well location map. *Source:* Redrawn from Kelkar, M. and Richmond, D. (1996) Implementation of Reservoir Management Plan – Self Unit, Glenn Pool Field. SPE/DOE 35407 – Society of Petroleum Engineers. (b) Increase in production in the Self Unit after the implementation of the reservoir management plan. *Source:* Redrawn from Kerr, D., Ye, L., Bahar, A., Kelkar, M. and Montgomery, S.L. (1999) Glenn Pool Field, Oklahoma: a case of improved production from a mature reservoir. E & P Notes AAPG Bulletin V.83 - 1 (January) 1-18. American Association of Petroleum Geologists.

perforated the reservoir resulting in an initial rapid increase in production followed by a steep decline (Kerr et al. 1999a). Cumulative production during 1906–1993 was 2.76 MMBO. In mid-1995, a reservoir management plan based on facies architecture, geostatistical simulation, flow simulation, and economic evaluation was partially implemented, which involved reperforating existing wells to expose more of the reservoir coupled with higher water injection rates. Substantial initial improvements in oil production rates were achieved, although water production remained high and production has since shown a decline, suggesting that injected waters are transferring preferentially through the most permeable reservoir sands (braided channel-fill facies). Future production is expected to come from meandering-channel (lateral accretion bar) and crevasse-splay sand facies (Kerr et al. 1999b).

8.6 Schiehallion Field, Faroe–Shetland Basin, West of Shetland

The Schiehallion Field is located in the Faroe–Shetland Basin, 200 km west of the Shetland Islands within the UKCS and in water depths of between 350 and 600 m. The field was discovered in 1993, appraised through 1994–1995, and development was sanctioned in 1996 with first oil produced in July 1998. At sanction, 935 MMbbls were identified within the closure, of which 340 MMbbls were expected to be recovered by the initial development: a recovery factor of 37%. As part of the reservoir characterization process a series of time-lapse seismic surveys were performed every three to four years resulting in the recognition of significant volumes of bypassed oil.

Hydrocarbons are trapped by a combination of stratigraphic and structural elements (Figure 8.7) in a series of stacked and isolated deepwater, submarine fan channels. The reservoir sands are part of the Vaila Formation of Paleocene age; the high-quality channel sands are named after a regional sequence stratigraphic scheme developed by the operator for the Paleocene–Early Eocene. The main producing channel sands have a permeability range of 800–1600 mD, with correspondingly good porosity, 25–30%, and net : gross ~70%; interchannel areas have lower net : gross values of 30–50%. Initial pressure measurements indicated that all reservoir zones were in communication; however, during

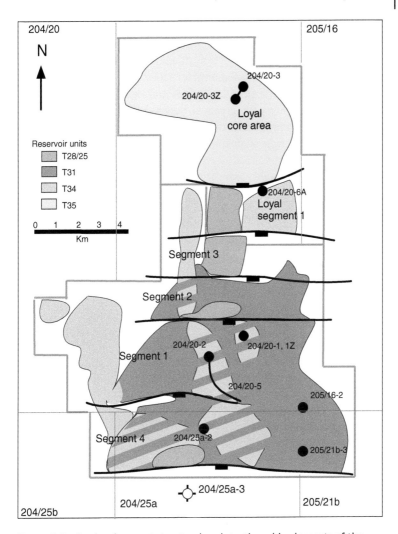

Figure 8.7 Predevelopment structural and stratigraphic elements of the Schiehallion Field, west of Shetland. *Source:* Redrawn from Freeman, P., Kelly, S., MacDonald, C., Millington, J. & Tothill, M. (2008) The Schiehallion Field: lessons learned modelling a complex deepwater turbidite. In: The Future of Geological Modelling in Hydrocarbon Development. Robinson, A., Griffiths, P., Price, S., Hegre, J. & Muggeridge, A. (eds) Geological Society Special Publication 309, 205-219.

extended well tests and early production, pressure depletion and weak injection support indicated that the reservoirs were not as well-connected as anticipated. Heterogeneity was attributed to crosscutting faults and facies variation at the edges of channels as they eroded into each other.

The oil has a gravity range of 22–28 °API and gas-oil ratio (GOR) 340 at initial pressure of 2150 psi; this is close to bubble point, indicating that water injection would be required from the start of production. At project sanction, 12 horizontal/high-angle production wells and 10 similar water injectors were planned; a gas disposal well was also included in the development. After some disappointing results, two further phases of drilling were initiated bringing the total number of producers by 2003 to 18, with an equal number of water injectors. Water breakthrough occurred in 2001, after less than three years of production. The wells were drilled through four subsea templates and connected to a re-purposed oil tanker as an FPSO via flexible risers.

During the first 15 years of production, significant upside was recognized in deeper reservoirs that became the target of a redevelopment program that began in 2012 and was completed in 2017. Remaining hydrocarbons are estimated at ~960 MMBO: near doubling the original in-place volume. Throughout the redevelopment, the field was shut-in, although water injection continued. A major drilling campaign was mounted to improve recovery from the field. Seventeen new producers and eight water injection wells were drilled; a further 24 infill wells can be accommodated by the redevelopment. The existing wells and manifolds will be tied back to the new FPSO via subsea flowlines and flexible risers. Of the 21 risers, 15 are for producers, 3 are for gas risers and 3 are for water injection. Seven new subsea manifolds and five drill centers have been installed, extending the capacity to beyond 75 wells. This is the world's largest sub-sea development.

During the shut-in, reservoir management activity continued. The period of field stability created an opportunity to update the full field dynamic model (FFM) of the reservoir and to complete a history match and reservoir uncertainty study. Multidisciplinary reviews were carried out to assess field panel performance and to identify remaining in-place hydrocarbons. Desk studies of any well performance issues led to the replacement of production/injector trees as well as a number of interventions to obtain surveillance data: pressure data, saturation log data, and production log data (PLT). The output from the simulation studies and panel reviews were used to identify infill opportunities.

The harsh weather and sea conditions meant that the original FPSO required replacement and this was completed with the installation of the new build Glen Lyon FPSO that was designed to handle 130 000 BOPD and 220 MMscf/d of gas. Claimed to be the world's largest harsh environment FPSO, Glen Lyon measures 270 m long by 52 m wide and weighs around 100 000 metric tons. It is anchored by a weathervane turret mooring system and features 20 mooring lines each of which are 1.6 km long. The success of the project will be measured through continued seismic survey, pressure monitoring, and further appraisal drilling. With the redevelopment, BP and partners expect to extend the field life to 2035 and beyond and recover an additional 450 MMBO. The produced oil is offloaded to a shuttle tanker for transport to the Sullom Voe oil terminal on the Shetland Islands for export. Schiehallion is jointly developed with the adjacent Loyal Field, which has the same operator.

8.7 North Burbank Field, Cherokee Basin, Oklahoma, USA

The North Burbank Field is located in the Cherokee Basin, Northeast Oklahoma, USA. The field was discovered and came onstream in 1920 and is part of a semicontinuous cluster of fields known as the Burbank Field Complex. In-place reserves are 671 MMBO and ultimate recoverable reserves are estimated as 358 MMBO. Several light oil pools are contained in a stratigraphic trap formed by updip pinch-out of fluvial channel sands, which incise into marine shales. The Middle Pennsylvanian Burbank Formation reservoir is locally divided into three subzones (Figure 8.8) by discontinuous shale layers and consists of very fine- to medium-grained litharenites deposited as part of a regional, south-draining fluvial system. Production was initially by solution gas drive, but the very low reservoir pressure necessitated the implementation of gas and water injection soon after production began, and these have ensured that reservoir pressure is maintained at or above initial pressure. Peak production was 122 000 BOPD in 1923, soon after start-up; however, this was not sustained, and production had declined to ~1600 BOPD by 2012. Several variably successful polymer/surfactant flooding projects have been undertaken from 1970 onward to reduce channeling of injected water. A program of CO_2 injection was initiated in 2013 to further enhance oil recovery.

The Burbank Field was developed rapidly following its discovery in 1920, and by 1924, 75% of the wells had been drilled in the main part of the field. The initial potential of wells varied from 10 000 to 12 000 BOPD

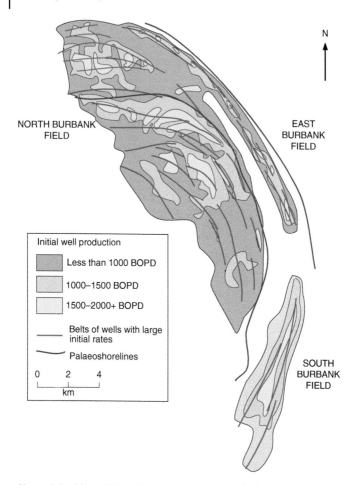

Figure 8.8   Map of North Burbank and adjacent fields showing the distribution of initial well production rates and locations of sweet spots. *Source:* Redrawn from Johnson, C.L. (1992) Burbank Field-USA, Anardarko Basin, Oklahoma. In: Stratigraphic Traps III (treatise of Petroleum Geology Atlas of Oil and Gas Fields) N.Foster (ed). American Association of Petroleum Geologists 1992, 333-345.

with an associated production of gas at an initial pressure of 800 psia. Little or no water was produced in the main part of the field until water flooding began. Peak production of 122 000 BOPD by solution gas drive was reached in 1923 from 1020 wells (Johnson 1992). Initial well production rates were mostly <500–1500 BOPD, with localized areas

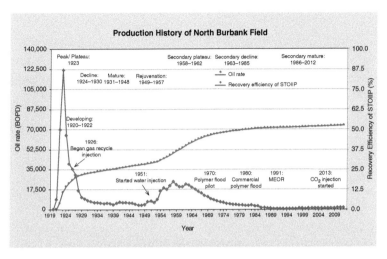

Figure 8.9 Production history of North Burbank Field showing significant events in reservoir performance and improved recovery methods.

producing >1500 BOPD (Figure 8.9). Pressure maintenance measures were initiated early on in the field life in order to prevent reservoir pressure falling below the bubble point. Produced gas was injected into the reservoir, commencing in 1926, and ~220 MMBO had been produced from the Burbank Field Complex by primary and gas injection methods before field-wide water injection was implemented in 1951 by the various operators (Johnson 1992). This was the largest water flooding project in the world at the time and remains the only field-wide secondary recovery.

Two factors influenced the efficient development of the Burbank Field; first, the Osage Indian Nation managed their assets well by requiring each 160 ac tract to be evenly drilled up before the adjacent tracks could be leased, and second, early in the field's development, operators realized the benefit of maintaining reservoir pressure, hence the early adoption of produced gas reinjection. The North Burbank Field Complex as a whole has seen a steady decline in production from just under 15 000 BOPD in 1970 to ~1600 in 2012.

A number of secondary and tertiary recovery strategies have been tested or implemented on various tracts of the field, including water injection, polymer and surfactant injection, microcellular injection, short radius drilling, and different stimulation methods.

8.7.1 Water Injection

Water injection was initiated in 1949 in the south of the field on a 90 ac pilot and nine infill injection wells were drilled to form nine five-spot patterns of 20 ac each. Poor sweep efficiency was recorded as a result of channeling of injected water along fractures. In 1954, the flood pattern was changed to a staggered line drive, with all injection wells drilled in E–W rows so that flooding occurred perpendicular to the predominant fracture system (Clampitt and Reid 1975).

Water injection in tract 49 in the north of the field began in 1959 and 60 days after the first oil response water-cut exceeded 80% and went on to reach 98%. Injected water was channeling through high permeability sands at the top of the reservoir and it was estimated that up to 90% of the injected water was flowing through only 15% of the reservoir thickness (Clampitt and Reid 1975). As part of the overall reservoir management process, wells with very high water-cut >98% were recompleted or shut-in.

8.7.2 Polymer Injection

A small polymer injection project was undertaken in tracts 40 and 49 in the northwest of the field in 1970 and produced incremental oil for 12 years, at an incremental rate of 50–60 BOPD. The project consisted of four injection wells in two rows offset by 12 producing wells aligned in three rows of four each. Injection of the polyacrylamide into the four injectors began in October 1970 and ended nine months later having injected ~85 000 lbs of material dissolved in 1.7 MMB of fresh water. The success of this small-scale pilot resulted in an expansion of the project in 1975 and again in 1980 in tract 97. A surfactant/polymer pilot was initiated in 1975 and a freshwater polymer injection project incorporating 36 wells and covering 1440 ac in the south-central part of the field was initiated in 1980 (Zornes et al. 1986). The freshwater polymer injection project led to a doubling of production rate and a decrease in the water–oil ratio.

On the whole, these projects have yielded incremental oil, but recovery has been about half that initially expected in places and has been adversely affected by unpredictable vertical and lateral permeability heterogeneities, high consumption of injected sulfonate due to adsorption on the oil-wet pore surfaces, channeling along fractures, and low fracture pressure (Lorenz et al. 1984; Szpakiewicz et al. 1987).

8.7.3 Short Radius Drilling

Drilling lateral wells from the same vertical borehole is a proven technique in many fields. A new well was drilled in the western part of the Burbank Field on the West Little Chief prospect and two laterals subsequently drilled for ~1000 ft to the east and west. The initial production was 320 BOPD and considered disappointing when compared with the expected rates of 2000–2500 BOPD. The reduced rate was put down to formation damage caused during drilling: xanthan gum was used in the drilling fluid to increase cuttings removal; however, it was suspected that this polymer, which has a great affinity for quartz grains, invaded the near borehole formation and reduced permeability. Remedial treatment using strong oxidizers was tried, as was an acid wash, but neither were effective. Even after placing the well on a pump, the best pump rate was 4 BOPD and 318 BWPD. Further attempts to drilling similar wells have not been attempted.

8.7.4 Microbial Selective Plugging

Microbial nutrients were injected in one injection well in the field to plug off high permeability layers and redirect injection water to lower permeability, higher oil-saturated zones. Several different types of treatment were tried using both sequential and co-injection of nutrients. Pressure fall-off/injection test and vertical injection profiles were conducted before and after nutrient injection. The results of pressure injection tests after co-injection of nutrients and an incubation period indicated a 33% drop in the effective permeability to the injection fluid and a negative skin factor; injection profile surveys showed a 33% reduction in the zone taking water. Swabbed samples from the well after shut-in revealed abundant bacterial concentrations and products as well as oil and inorganic solids. Although permeability reduction was observed, the unstable behavior of the plug suggested that insufficient biomass was being formed to effectively seal off the high permeability zones.

8.7.5 Downhole Vibration Stimulation

An experimental downhole vibration tool (DHVT) was developed to be installed in 7 ins casing for a field test of vibration stimulation in a mature water flooded field and to evaluate the effects of vibration on both produced fluid characteristics and injection well performance: downhole geophones and hydrophones were used to monitor the downhole vibrations. However, during the first week of a 90-day field test the DHVT became stuck during routine retrieval activity; the tool remains stuck in the hole and the field test was ended. No response to the stimulation was observed to the produced fluids or the injection characteristics or the wells monitored.

8.7.6   Carbon Dioxide Injection and Sequestration

The current operator, Chaparral Energy, began a tertiary recovery program of CO_2 injection in 2013 that may recover an additional 88 MMBO over the next 30 years and increase daily production to ~7500 BOPD by 2018. The plan includes a ~68 mi, 8 in pipeline from an existing fertilizer plant in Coffeyville, Kansas, together with a 23 500 hp compressor station designed to move up to 60 MMscf/d of captured CO_2 (Oil and Gas Journal 2013).

8.8   Nakhla Field, Hameimat Trough, East Sirt Basin, Libya

The Nakhla Field is located onshore in Libya in the Hameimat Trough, East Sirt Basin. The field was discovered in 1970 and came onstream in 1995. It has STOIIP of 933 MMBO. Light oil is contained in a tilted fault block which formed mainly during Cretaceous rifting and was modified by later dextral wrenching (Figure 8.10). The Barremian, Upper Sarir Sandstone Formation reservoir was deposited in a lacustrine fan delta environment. Fluvial, sheetflood, and interchannel sands with jigsaw puzzle architecture dominate the reservoir interval. Porosity is 10–14% and permeability is low at 0.1–20 mD. Reservoir quality is largely facies-dependent, with higher-energy fluvial sediments forming the best reservoirs. Fracturing locally enhances permeability. Barriers and baffles to flow include faults, lacustrine shales, volcaniclastic and basaltic layers, and diagenetically-altered fracture zones. Nakhla oil is 42–43 °API and has a GOR of 1250 SCF/STB and production takes place through solution gas drive. The best producers are vertical wells that have been fracture stimulated. Pressure maintenance by water injection is unsuitable at

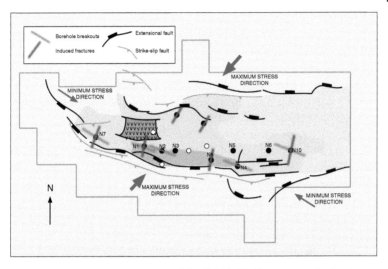

Figure 8.10 Depth structure map of the Nahkla Field top reservoir showing the horst bound easterly faults. Also depicted are the orientation of borehole breakouts and induced fractures. *Source:* Redrawn from Droegmuellar, U. and Leonhardt, B. (2005) Hydraulic Frac Stimulations ina Libyan Oil Field – A Case History. SPE 95019 – Society of Petroleum Engineers.

Nakhla. A single horizontal well was drilled but performed no better than the vertical wells. Peak average production of 10 730 BOPD was attained in 1997. Average production for 2004 stood at 8579 BOPD, and by end-2004, cumulative production was 28.6 MMBO.

The field came onstream following completion of a dedicated gas–oil separation plant (GOSP). Well 1, which had actually been drilled in 1993, tested at 2900 BOPD but was shut-in until 1995, when the GOSP was constructed. Of the two remaining wells, drilled as part of Phase 1, one encountered volcanic rock and the other suffered technical problems before encountering reservoir (Droegemueller and Leonhardt 2005).

Ten further wells were drilled during the second development phase from 2002 to 2004, with limited success. The first of these (well 9) was a horizontal well with a 3000 ft open hole section completed in the Sarir Sandstone. Well 10 encountered the OWC for the first time in the east of the field, but the depth of the contact is not published. The well was not considered a candidate for fracturing due to its proximity to the water column. Well 7 encountered oil in the west of the field in an extremely low (0.1 mD) permeability part of the reservoir. Of the remaining seven wells, two encountered volcaniclastics, three penetrated tight reservoirs, and one was successfully fractured (well 8) in 2004 (Droegemueller and Leonhardt 2005).

The operator implemented an integrated development plan for all of the C97 concession in 2004, which began with centralization of three existing oil-processing facilities into one upgraded unit at Nakhla. Throughput capacity was increased from 15 MBOPD to 30 MBOPD and from 15 MMCFGPD to 40 MMCFGPD and the installation of multiphase pumps and flowlines between the other fields on the concession and Nakhla was begun. The oil is processed after separation with desalters and dehydrators. Due to the high pour point of the oil and to avoid wax precipitation, a hot water circulation system was installed. The produced oil is measured in a metering station and exported by means of an existing export pipeline to the Ras Lanuf terminal on the Mediterranean coast.

The start of production was in 1995 at a rate of ~5500 BOPD, which increased to a peak of ~10 800 BOPD when all the Phase I wells came onstream after fracture stimulation. Thereafter production declined at ~15% per year until Phase II drilling and production commenced in 2002, boosting production to ~8800 BOPD, and subsequently further declining to ~8500 BOPD in 2004. The field has reached a mature or declining stage of development.

Well productivity appears to vary randomly across the field, both pre- and postfracture stimulation, although all wells show a clear short-term (six months) improvement in performance following fracturing. The maximum productivity index was 4.27 BOPD psi^{-1}. Longer-term comparison of the effectiveness of fracturing, using decline curve analysis to predict rates without stimulation, confirms the improvement in the rate of production. Additional recovery is also predicted, attributable to well bottomhole pressures remaining above bubble point for longer once fracturing has taken place. The field had a cumulative production of 28.6 MMBO by November 2004 and production rate throughout 2004 was ~8579 BOPD (Figure 8.11) (Droegemueller and Leonhardt 2005).

A pattern of decreasing permeability as production takes place has emerged which is attributed to one or more of the following: (i) severe reservoir heterogeneity, whereby early well tests record production preferentially from high permeability layers; (ii) depletion of reservoir

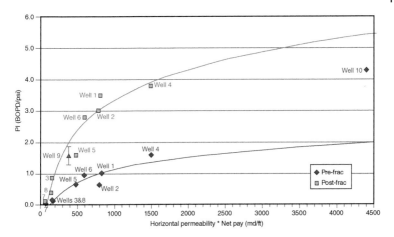

Figure 8.11 Postfracture stimulation production improvement seen in selected wells after a six-month period. Productivity of the horizontal well 9 was no more cost-effective than fracture stimulation. *Source:* Redrawn from Droegmuellar, U. and Leonhardt, B. (2005) Hydraulic Frac Stimulations ina Libyan Oil Field – A Case History. SPE 95019 – Society of Petroleum Engineers.

pressure and concomitant compaction; and (iii) reduction in relative permeability due to production below bubble point pressure.

Gas injection has been investigated as a possible improved recovery method for the Nakhla Field, as water injectivities are very low even in fracture-stimulated wells and a prohibitive number of water injectors would be needed (Beuthan et al. 2006). Miscible and immiscible gas injection techniques were being investigated in 2006, although it is not known whether IOR has yet been implemented.

In the first development phase, a program of fracture stimulation was implemented to overcome the poor reservoir quality. Well 1 was fractured in 1995, followed by wells 2 and 3 in 1996, and wells 4, 5, and 6 during 1997. Fracture stimulation produced a threefold increase in productivity in some wells. The productivity of the completed horizontal section in well 9 was comparable to the previous fractured vertical wells and, as such, subsequent wells were vertical as this was more cost-effective.

Overall, 80% of the production wells in the field have been successfully stimulated by hydraulic fracturing treatments. State-of-the-art real-time analysis was performed to evaluate and optimize the fracturing operations to compare pre-frac and post-frac well test results obtained from pressure build analysis (Droegemueller and Leonhardt 2005). To monitor potential aquifer influx, thermal decay time (TDT) logging was carried out in well 10 one year after coming on production: no change in water saturation behind pipe was observed, supporting the assumption of a weak or inactive aquifer.

8.9 Forties Field, Central North Sea, UKCS

The Forties Field is located in blocks 21/10 and 22/6a of the UK North Sea, ~180 km ENE of Aberdeen. BP drilled the discovery well, 22/10-1, in 1970 with the southeasterly extension into block 22/6a proven by well 22/6-1 drilled by Shell/Esso in 1971. The reservoir comprises Paleocene submarine fan sandstones deposited mainly by high-density turbidites within multiple channel/lobe systems (Figure 8.12). Hydrocarbons are trapped in a four-way dip-closed anticline draped over the Forties–Montrose High. Laterally extensive interchannel and abandonment mudstones, and minor synsedimentary faulting, compartmentalize the field. Three main compartments are recognized: Main Sand, the Charlie Sand and SE Forties. The field has an active aquifer, and pressure support is also provided by peripheral water injection. The field has been

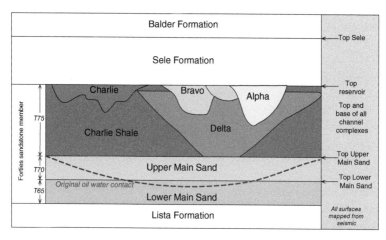

Figure 8.12 Main stratigraphic units of the Forties Field based on well penetrations and seismic interpretations. *Source:* Vaughan et al. (2007): modified and redrawn Redrawn from Vaughan, O., Jones, R. and Plahn S. (2007) Reservoir Management aspects of the Rejuvenation of the Forties Field, UKCS. SPE 109012 – Society of Petroleum Engineers.

developed with four platforms on the main accumulation and a fifth to exploit the southeastern part of the field. The field came onstream in 1975 and plateau production was reached in 1978–1980 at ~500 000 BOPD. The oil is exported via a dedicated Forties pipeline system (FPS) to St. Fergus on the Scottish mainland. Initial oil in-place was estimated at 4200 MMBO with an expected ultimate recovery of ~59%. Since 2003, the field has been operated by Apache Corporation who have implemented a major redevelopment of the field including drilling over 100 infill targets, a series of 4D seismic surveys, and an additional platform to maximize the field recovery. To date (2020), 2800 MMBO have been produced, representing ~65% recovery, and an increase of ~230 MMBO over the original estimate.

Four fixed-steel drilling/production platforms (Alpha, Bravo, Charlie, and Delta) (Figure 8.13) were installed on block 21/10 in 1974–1975, and the field was put onstream through Forties Alpha in September 1975. Conventional, near-vertical production wells were targeted at the high net:gross (NTG) channel sandstones and were highly productive, with just 16 wells producing the first BBO (Leonard et al. 2000). Soon after

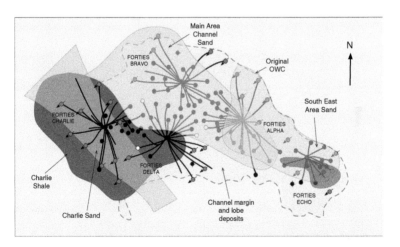

Figure 8.13   Forties Field platform and well locations with distribution of major sand and shale bodies identified. *Source:* Smith, L., and Perras, L. (1998) The Brimmond reservoir extending the Forties Field into the Millennium by maximising profits without expensive rig upgrades. SPE 50383 In : Proceedings SPE International conference on Horizontal Well Technology, Calgary. Society of Petroleum Engineers. and (Kulpecz, A.A., and van Geuns, L.C. (1988) Forties Field (UK) - geological modelling of a turbidite sequence for reservoir simulation. AAPG Bulletin v. 72, no. 2, 209. American Association of Petroleum Geologists.

production began, it became apparent that the Charlie Sand in the western part of the field was depleting more rapidly than the rest of the field and seawater injection was initiated in 1976. Reservoir pressure across the field fell by 800–1000 psi during the first seven years of production but was restored to <100 psi below the initial pressure by 1995. Water injection resulted in excellent vertical and areal sweep across most of the field, giving residual oil saturations of 15–30%. Oil recovery in areas of the field isolated from water injection was significantly reduced (Leonard et al. 2000). Several other improved oil recovery schemes were examined during the life of the field, but none were implemented.

Plateau production of ~500 000 BOPD was reached in 1978–1980, with well rates of 3000–24 000 BOPD, averaging 11 000 BOPD (Figure 8.14) (Wills 1991). Water production began in 1978 and was often associated with increasing sand production because of the friable nature of the reservoir. Barium sulfate scale deposition due to incompatible formation and injection waters also caused problems (Brand et al. 1995; Jones et al. 1997). Water coning, however, was less problematic than expected, owing to the very low $k_v : k_h$ ratios of the reservoir sands and the numerous thin shale layers (Langley 1987). In 1985, development drilling in

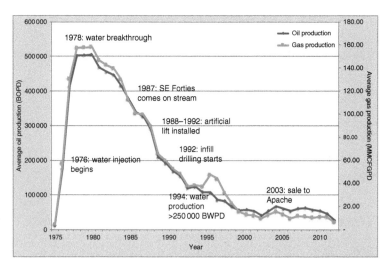

Figure 8.14 Forties Field oil and gas production from 1975 to 2015, showing major interventions and events. *Source:* Newly added source. Tag accordingly.

block 22/6a began from the Echo platform at SE Forties, which was put onstream in 1987 (Kulpecz and van Geuns 1988). Artificial lift began with the installation of electric submersible pumps in the wells at SE Forties in 1988, and the installation of gas lift facilities on the four platforms in the main field area during 1989–1992 (Wills and Peattie 1990; Brand et al. 1995). This stabilized the falling production rate at ~110 000 BOPD during 1993–1996, before it declined again to ~42 000 BOPD by 2003. Water-cut was 69% in 1995 (Brand et al. 1995).

After the high N:G channel facies had been swept relatively successfully by a combination of strong aquifer drive and water injection, a program of infill drilling began in September 1992 to access bypassed oil in lower N:G areas, marginal to the channels and toward the channel tops (Leonard et al. 2000). Infill targets were initially identified using the 1988 3D seismic time-lapse data, but further 3D data were acquired in 1996 and 2000 to allow new seismic imaging of lithology, fluids, and saturation changes across the field. By 2000, 65 wells had been drilled, including three infill injectors, 37 failed well replacements, and 25 infill producers that produced 160 MMBO (Jones et al. 1997; Leonard et al. 2000). Multilateral and horizontal wells were drilled in 1997 but did not deliver significantly better targets than previous wells. In 1999, through tubing rotary drilling (TTRD) was implemented in an attempt to significantly reduce drilling costs, and two TTRD wells were completed on Forties Bravo (FB) (Carter and Heale 2003).

BP sold its 96.14% stake in the Forties Field to Apache North Sea Ltd. in January 2003. At that time Forties was a giant field in decline: production had fallen to an average rate of 41 000 BOPD and was declining. Total production at time of sale was 2.5 BBO, with estimated remaining reserves of 144 MMBO. No new wells had been drilled since 2001 when the planned infill campaign was cancelled. In the following eight years, Apache drilled over 100 infill targets at an overall success rate of 74%, translating to over 80 production completions and 4 injection completions. As a result of the drilling campaign, the annualized average production rate has been held at ~60 000 BOPD since 2005, and cumulative production in the same period was over 150 MMBO, of which half is from the new wells.

The best wells, with initial production rates of 3000–6000 BOPD, were targeted at strong oil anomalies on the 4D data and indications of a lack of sweep on the 4D difference volume (Rose et al. 2011). A further element in the evaluation of bypassed pay traps is the identification and mapping of the many small discontinuities present in the seismic data. These features, derived from seismic coherency and spectral decomposition datasets, are associated with sand-body edges and small faults with 5–10 m throw; such edges are commonly seen to line up with the edge of a swept zone imaged by 4D seismic. The difference between the seismic response of brine and oil-filled sand has also been exploited by the use of 4D seismic. Difference volumes between each successive 3D survey show a well-developed response to water replacing hydrocarbons that can be related to particular production wells. Areas that the seismic indicates remain unswept become the future infill targets if supported by the local distribution of seismic discontinuities, historic production offtake, and sand distribution. There is a dynamic and constantly evolving remaining target portfolio, the strength of which has allowed a new 18-slot satellite platform to be installed in 2012.

Between 2003 and 2008, Apache invested $1.24 billion of capital in the rehabilitation of the Forties Field, much of which was allocated to infrastructure improvements and drilling (Vaughan et al. 2007). Originally, each platform had full separation and export pumping facilities and was connected directly to the FPS. From the mid-1990s, processed crude was sent from each platform at reduced pressure to Forties Charlie (FC) where it was combined and boosted in pressure for export. A significant aspect of the Apache redevelopment has been to upgrade or replace unreliable facilities in support of a longer-term view of remaining field life. Electrical submersible pumps (ESPs) have been introduced to FB wells with the goal of attaining a not normally manned (NNM) status, with production of well-stream fluids directly to FC for separation. Ten separate power generators have been replaced with two turbines on each of FA and FC, with power and fuel gas now being redistributed through the field as necessary via a new set of subsea power and gas lines. This upgrade has increased power capacity and reliability and reduced fuel costs and flaring of unused gas.

To achieve the planned production increase required the reinstatement of drilling capability on all four main platforms: all were fitted with top-drives and iron roughnecks, and two with oil-based mud handling and cuttings reinjection systems. The newly-defined infill targets require the drilling of higher-angled wells that are threaded through stacked thinner-bedded oil-bearing sands sandwiched between thicker swept zones. Cased completions are also implemented to isolate short sections of water-bearing sands. The friable nature of the sands has resulted in sand production issues over time as the reservoir pressure has declined; cased and perforated completions have also proved the most reliable way to control the sand influx, although other completion types have been used including screens (Vaughan et al. 2007).

Real-time LWD has been used continuously in drilling new wells. Typically, producers are drilled with GR, resistivity, and nuclear tools in the BHA through the reservoir section to evaluate whether sufficient pay has been penetrated to warrant a completion; if so, then a baseline cased hole saturation log is run in the liner. Azimuthal LWD density images have been used to identify thinner-bedded pay, which has become an important target in some parts of the reservoir. Shear sonic tools have been run for seismic calibration as well as NMR for movable hydrocarbon evaluation. The cost of LWD logging is closely monitored for each well to balance this against the value of information gained. Open hole wireline logs are considered largely redundant because of the well geometries (Vaughan et al. 2007).

Many of the technologies considered to improve recovery are now standard practice in Forties, such as 4D seismic, high-resolution biostratigraphy, static and dynamic modeling, produced water reinjection, artificial lift methods, and extended reach drilling. However, in 1984 studies were conducted that showed that both miscible and immiscible gas injection schemes were not economically viable at the time, and neither were polymer or surfactant floods. Today, however, with rising oil prices and new technologies, these and other methods may once again be considered.

8.10 Leman Field, Southern North Sea, UKCS

The Leman Field was discovered in 1966 in the Sole Pit Basin, part of the UK sector of the Southern North Sea in water depths ~120 ft. The GIIP is estimated as 14 TCFG, with ultimate recoverable reserves of 12.7 TCFG, giving a recovery factor of 91%. The reservoir comprises Upper Permian Rotliegend sandstones of the Leman Sandstone Formation deposited as large-scale aeolian dunes. Gas is trapped in a broad anticline that is primarily dip-closed but compartmentalized by numerous faults

(Figure 8.15). The field has an inactive aquifer so produces under natural depletion (Hillier 1990).

The original development plan for the field proposed a tank-like reservoir, needing only a limited number of crestal drainage points. However, widespread fault compartmentalization and the relatively poor recovery from the low-permeability zone A, meant that a wider spread of production wells was needed (Hillier and Williams 1991; Hillier 2003). By 2003, 192 production wells had been drilled from 16 platforms (Hillier 2003). Perforated reservoir intervals are up to 700 ft thick (Dyck and Moore 1994). By the early 1990s, the more permeable zone B showed substantial pressure depletion, with reservoir pressures as low as 300 psi (compared to a slightly above hydrostatic initial reservoir pressure of 3022 psi) in the

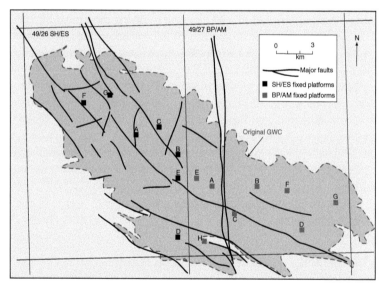

Figure 8.15   Leman Field outline with major structural features highlighted together with platforms operated by Shell/Esso and BP/Amoco in 1991. *Source:* Hillier, A.P., and Williams, B.P.J. (1991) The Leman Field, Blocks 49/26, 49/27, 49/28, 53/1, 53/2, UK North Sea. In: United Kingdom Oil and Gas Fields, 25 Years Commemorative Volume: Abbotts, I.L. (ed.) Geological Society, London, Memoir, no. 14, p. 451-458. and Hillier, A.P. (2003) The Leman Field, Blocks 49/26, 49/27, 49/28, 53/1, 53/2, UK North Sea, in Gluyas, J.G., and Hichens, H.M., 2003, eds., United Kingdom Oil and Gas Fields, Commemorative Millennium Volume: Geological Society, London, Memoir, no. 20, p. 761-770.

crestal and southeastern parts of the field, where the original GWC lies within or below zone B. In areas where zone B lies below the GWC, such as the northwestern part of the field, near virgin pressures were locally encountered (Dyck and Moore 1994).

Gas production began in August 1968 and reached a peak of 1761 MMCFGPD in 1976, thereafter declining steadily to ~340 MMCFGPD in 1994, and then maintaining ~300–400 MMCFGPD through to 2004, when the field had produced 11.1 TCFG. Since then, production rates have fallen to 100–150 MMCFGPD and the total gas produced to date (2012) is 11.6 TCFG (Figure 8.16). It was estimated that field life would extend to at least the year 2011 (Dyck and Moore 1994) and this is proving to be the case with new step-out wells being proposed and drilled. The gas is transported via two 30 in pipelines to the onshore UK terminal at Bacton (Hillier 2003). Originally, there were three pipelines, but one was decommissioned in the mid-1990s as part of the rationalization of the Shell/Esso unit that also saw a reduction in the number of platforms and their de-manning.

In the low-permeability areas of the field, massive hydraulic fracture treatments are necessary to achieve commercial production (Clayton and Gordon 1990). The northwestern part of the field was developed in 1987–1990 by 14 production wells. A fracturing campaign (involving

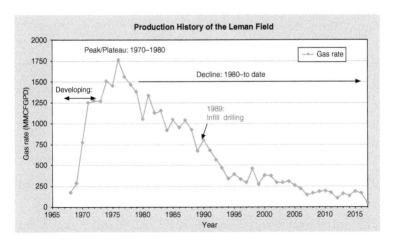

Figure 8.16 Production history of the Leman Field with typical life cycle from development through plateau and into decline. Prior to 1995, gas production was contracted annually rather than daily after the market was liberated.

injection of ~10 million lbs of sand proppant, initially placed using a diesel-polymer emulsion fracture fluid, and, after 1988, a water-based cross-linked gel fluid), resulted in well flow rates of 5–40 MMCFGPD, averaging 13 MMCFGPD. These rates were better than expected and were considered to be the result of the induced fractures intersecting a natural fissure system (Clayton and Gordon 1990). Fracture stimulation led to increased water production and also high levels of proppant backflow, which sometimes caused a significant reduction in gas production (Dyck and Moore 1994).

Compressors have been progressively introduced to the field and it is expected that by the end of field life, gas will be produced at suction pressures of 30 psi. Eliminating flow restrictions has increased the gas-producing capacity of existing wells, but the falling reservoir pressures allowed condensate and formation water to build up in the well bore, significantly impairing production performance. This is primarily prevented by periodically shutting-in wells to allow pressures to recover above critical limits for each well (Dyck and Moore 1994).

The operator of the eastern producing unit, Perenco, decided to develop a new tieback to the Leman 27A platform by means of an 8 ins diameter, 9.2 km pipeline. Perenco drilled and tested a new step-out well (53/02-14A) in 2011 with the intention of producing gas from the south-eastern margins of the field. The platform was installed in 2012 and is located in block 49/27, potentially providing a hub for further tieback projects. The pipeline will be equipped with an umbilical piggyback line for the transportation of MEG (mono-ethylene glycol) and housing of the hydraulic and electrical systems for the well control from the platform. The development went into production in 2013 and the average gas production is ~10 MMCFGPD. Condensate production is estimated to be ~88 BCPD. In 2019, plans to extend the life of this field along with Indefatigable ('Inde') gas field were approved. The Southern Hub Asset Rationalisation Project (Sharp) will combine the two 'oversized' Leman and Inde fields into one fit-for-purpose production hub. The operator said they hoped to extend the life of both fields and to reduce costs.

8.11 Summary

From the tales recounted in this chapter is it obvious that this industry is amongst the most innovative and rewarding for those employed in the many and varied roles.

Appendix 1

Guide to Reservoir Simulation

In 2004, I was incapacitated with a broken ankle: it was known as Steve's Self-Induced Injury. At the time, I was looking for a project to do at home and came up with the idea of an introduction to reservoir simulation for nonspecialists. I put together a course that covered building a simulation model, an introduction to special core analysis, and to petroleum economics. In this appendix, I plan to cover some of those aspects in greater detail than in the main body of the text; however, please do not think that this is a substitute for a detailed course in reservoir simulation or one of the many excellent papers or monographs published by the Society of Petroleum Engineers. I found particularly useful a book by Mike Carlson (2003) called *Practical Reservoir Simulation*.

This guide will follow the framework of a reservoir simulation study from setting objectives, sensitivity studies, data gathering, model construction, history matching, and prediction runs to reporting. Two areas not covered in the main book are the use of special core analysis, relative permeability measurements and pressure–volume–temperature (PVT) data. The following are key rules for any simulation study:

- Never start a simulation study without clear objectives.
- Determine which model parameters have the strongest influence on the study outcome: does the accuracy with which they are measured justify the study?

Reservoir Management: A Practical Guide, First Edition. Steve Cannon.
© 2021 John Wiley & Sons Ltd. Published 2021 by John Wiley & Sons Ltd.

- Take care with prediction results if they are reliant on parameters that are not validated in the history match.
- Wherever possible, use conceptual and analytical models, elemental studies, well models, and coarse grid prototypes before setting up the full-field model (FFM).
- Take care using 2D models because they do not capture out-of-plane flow paths leading to incorrect fluid phase distributions and well spacing.

A.1 Phases of a Reservoir Simulation Study

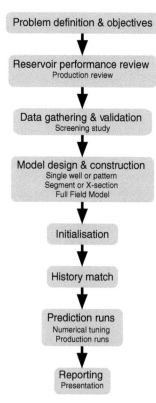

Figure A.1 Typical simulation study workflow.

Figure A.1 describes the nine stages of a reservoir simulation from problem definition to reporting; there will be some projects that require fewer steps and other that require more: this is just a way of making sure that the key elements of a project are considered in the workflow.

A.1.1 Problem Definition

The first step is to define the nature of the problem and to specify a likely solution: should you build a FFM to address the ultimate recovery, or a well model to look at production performance issues or well test evaluation. The use of analytical methods can often reduce the scope of the problem and give information on fluid displacement in the reservoir. At this stage it is important to gather all available data on the field and reservoir and not to discount any information as unreliable.

A.1.2 Setting Objectives and Terms of Reference

Do not start a study without a clear set of objectives agreed by all the stakeholders; it is worthwhile setting any history matching criteria at the stage. This is the point in the planning process to decide whether you have sufficient data to achieve the stated objectives, and to estimate the time and resources required. Often, a simple solution can be found using analytical tools.

A.1.3 Data Collection

Do not omit any data yet but determine whether there are any deficiencies (uncertainties) in the data set.

A.1.4 Data Validation

Relative permeability and PVT data usually need validation because in a large field study experiments may have been carried out in different laboratories or using different methods. All analytical data should be normalized; this includes capillary pressure and relative permeability data. Reconcile laboratory data with normalized wireline log measurements of rock properties.

A.1.5 Data Interpretation

The geoscience data should provide a detailed layering scheme and fault network with which to create the framework model. A description of the depositional model and facies should give the orientation of major flow units as defined by the petrophysical properties. If a 3D geological model has not been built, then interpret the available geological, petrophysical, and production data. Again, a decision should be made as to whether the data available justifies a simulation study, in particular if those parameters that have the strongest influence on the study outcome have not or cannot be measured accurately enough.

A.1.6 Model Construction

This might be as simple as upscaling an existing geocellular model, but it is also advisable to create segment models to test particular sensitivities such as fault transmissibility or grid block orientation. Based on such work, the grid size and layering may be simplified or varied in different parts of the model. The displacement process should be well understood at this stage and rock relative permeability known so that pseudoization of capillary pressure curves and relative permeability data can be generated. A set of PVT data and fluid properties is required, as well as production history data and other reservoir monitoring data.

At this stage a FFM will exist and an in-place hydrocarbon estimate determined. This is the obvious point at which to gather the stakeholders and agree the approach and results to date. These steps are often the most time-consuming and it is not unusual to have burned two-thirds of the budget by now.

A.1.7   History Matching

History matching may be straightforward if all you are looking for is a pressure match for a gas depletion study or undersaturated oil production, but increasingly more complex objectives are being asked for, such as water or gas breakthrough times and GOR trends that require matching. It is during history matching that the stability of the model is discovered and run times are evaluated; it is now that numerical tuning and diagnostic action take place, in other words, the reservoir engineer 'tweaks' the model to get a smoother, faster, more 'reliable' outcome, often at the expense of the geological data. Use of volume or permeability multipliers are the go-to solutions for a poor history match. Often it is easy to match a FFM but individual wells or fault segments remain problematic, so a 'fudge factor' is introduced. There is normally a geological reason such as lower net:gross (NTG) or fractures that might influence the match, so talk to the geologist! Modern integrated computing tools allow for changes to be made to a model with relative ease.

A.1.8   Prediction Runs

These will follow on from the study objectives but be aware that although repeatable they may not be unique.

A.1.9   Reporting

Please always write a report of your study to include the objectives, methods, and results so that others can work with an existing model rather than reinvent the wheel, although this happens time and again. A PowerPoint presentation is not good enough unless you include a dialogue as well! A good report also forms the basis of a postmortem of the study when, in hindsight, it can be seen if and where wrong assumptions were made.

A.2   Data Gathering

Much of the data required for a simulation model comes from well logs and cores including porosity, permeability, and NTG. These data give us the input for resource estimation as well as production potential, together with well test data. It is important to reconcile these different datasets as they are representative of different scales of investigation and only provide a sparse coverage of a field or reservoir. The two main static properties are porosity and NTG and are generally straightforward to input into the simulation.

Porosity is the fraction of bulk reservoir rock that can contain fluids. The product of porosity and bulk volume gives pore volume, which is the default input for most reservoir simulators. Porosity is obtained from laboratory measurements and well logs. As they each are derived from different sources they should be compared for consistency: core data is considered more accurate, but wireline data is more continuous over a reservoir interval and represents an in situ measurement.

NTG is a measure of the possible porous intervals within the reservoir as opposed to non-reservoir. This is often treated as a cut-off based on porosity but, increasingly, a good facies model or total property method is used at least at the finer scale. Large non-reservoir intervals in the simulation model are seen as voided cells in the grid: 'numed out' but not set to zero. Smaller non-reservoir sections, usually less than a grid cell dimension, are dealt with by a pore volume multiplier to reduce the proportion of reservoir.

Permeability is a measure of the fluid conductivity of the reservoir rock and is therefore a dynamic property that also has directionality. Permeability can be measured in a number of ways. Most measurements are made on core data in the laboratory, but the use of nuclear magnetic resonance (NMR) in the well bore has been shown to give reliable indication of fluid movement in situ rather than at ambient conditions or in a pressure cell: these measurements still need to be calibrated with laboratory tests. Another phenomenon that has to be accounted for is the slippage of gas along the pore walls at low pressure; the so-called Klinkenberg effect.

$$k_a = k_k \left(1 + \frac{k_k - 0.33}{P} \right)$$

(Permeability is in mD and P is the average pressure in the core in atmospheres.)

Permeability is overestimated when measured by air through a core plug at low pressure. This effect is most pronounced in poorly consolidated reservoirs. Other direct measurements of permeability are through wireline formation tests, pressure transient well tests, and interference tests, which record well flow at the reservoir scale. It is important that all methods of determining permeability are compared as the property is also scale dependent.

Permeability is also a function of grain size and is often related to porosity through a semilogarithmic relationship. This method will often help define different rock types that can be used to characterize the reservoir. One common source of inaccuracy is that core permeability is compared with log-derived porosity.

Another relationship that has a large influence on the behavior of the simulation model is that between vertical and horizontal permeability. Once again, core measurements provide most types of k_v/k_h relationships either through mini-permeametry, pairs of plugs or triaxial plugs, and, finally, whole core samples. Each of the methods have limitations mainly due to sample size or, in the case of mini-permeametry, difficulties calibrating with the other methods. Ultimately, any of these methods still need to be upscaled for use in the average simulation grid cell ($250\,m \times 250\,m \times 10\,m$).

Capillary pressure in a porous medium is the pressure difference across the interface of two immiscible fluids that is caused by interfacial tension. Capillary pressure is a function then of the fluids, the interfacial tension, pore size distribution, and wettability, and is thus different for different rock types: a function of lithology, porosity, and permeability. Capillary pressure (P_c) is expressed in terms of the interfacial tension between the wetting and non-wetting fluid phases, σ, and the contact angle between the wetting phase and the rock surface, θ, as follows:

$$P_c = \frac{2\sigma\cos\theta}{r}$$

where r is the effective pore radius.

Capillary pressure can be measured in the laboratory on cleaned and brine-saturated core samples using a centrifuge and porous plate or a diaphragm method. Common fluid combinations used are air–brine, where brine is the wetting phase; oil–brine where oil is the displacing fluid; water–toluene; water–kerosene; and, finally, air–mercury: the latter experiment involves mercury being injected into the dried and cleaned sample. Each of these laboratory tests are subject to experimental errors: the sample may not be properly cleaned and dried or be fractured or the experiment may be terminated before complete saturation is achieved especially in low permeability samples. Most measurements are made during the drainage phase when the 100% wetting phase saturation is achieved, and the end point is positive or zero. The imbibition curve is seldom measured but would end at the residual non-wetting saturation. In most simulation studies, it is the drainage capillary curve that is used for model initialization because the migration of oil or gas into the reservoir is assumed to be a drainage process: hydrocarbon displacing water.

In the laboratory, the fluids involved are either water and mercury or simulated oil and brine depending on the experimental method, and these need to be converted to reservoir conditions using the following equation and conversion values presented in Table A.1.

$$P_{cRes} = \frac{P_{cLab}\left(\cos\right)_{Res}}{\left(\cos\right)_{Lab}}$$

The height-related capillary pressure data is related to a saturation-height relationship using the following equation:

$$H = \frac{P_{cRes}}{g\left(\rho_1 - \rho_2\right)}$$

where g is the gravitational constant and ρ_1 and ρ_2 are the densities of water and hydrocarbon respectively. Capillary pressure is a function of pore throat size, rather than pore volume, and is therefore subject to the effects of confining pressure at reservoir conditions. Care should always be taken when building a database of capillary pressure data to ensure consistency in experimental methods and conditions; different laboratories may use different techniques or equipment resulting in a systematic error (Table A.1) Saturation profiles can be obtained from well logs and converted to a saturation-height profile using the free water level (FWL) as the zero datum. Before being able to develop a log-derived saturation-height relationship it is necessary to convert measured depth values to true vertical depth subsea (TVDSS). This is best done using a directional survey and the appropriate algorithm in the log analysis package. The log saturation data should only include good quality data from the cleanest and thickest sands to eliminate uncertainty over clay-bound water and shoulder effects of the input resistivity logs. In most dynamic simulations the asymptotic water saturation of the capillary curve, commonly called the connate water saturation, should be equal to the irreducible saturation of the oil–water relative permeability curve.

Worthington (2002) identifies three categories of saturation-height relationship: single- and multi-predictor algorithms and normalized functions. Ideally, each rock type should have a unique saturation-height relationship based on either geological or petrophysical properties.

Table A.1 Conversion of laboratory capillary pressure data to reservoir conditions.

System	Contact angle θ (degrees)	Interfacial tension σ (dynes cm^{-1})
Laboratory conditions		
Air–water	0	72
Oil–water	30	48
Mercury–air	140	480
Generic reservoir conditions		
Gas–water	0	50
Oil–water	30	30

Cannon (1994) coined the term 'petrofacies' to establish a link between geological and petrophysical attributes of a unique, log-derived facies predictor, characterized by definitive mean values of porosity, permeability, and water saturation. Depending on the data available it is possible to generate different saturation-height relationships.

In the dynamic simulation, modeling saturation functions can be more challenging as, typically, several functions are used to define different reservoir zones or rock types. Commonly, one set of normalized curves having different end points is used to represent the variations. The starting point should be the facies model as this will better represent the petrophysical properties in the model.

The Leverett J-function (Leverett 1941) is an example of a normalized saturation function that relates porosity and permeability to water saturation through the following equation:

$$J\left(Sw\right)=\frac{Pc}{\sigma\cos\theta}\sqrt{\frac{k}{\phi}}$$

where P_c is the pressure differential between the FWL and the measured point ($P_c = gH(\rho_1 - \rho_2)$) and the $\sigma\cos\theta$ term is the surface tension and contact angle from laboratory experiments. When using log-derived saturation data as input it is not necessary to include the contact angle and viscosity terms as the data is already at reservoir conditions, so only height and fluid density are required as additional input. When plotting saturation against height for different rock types it becomes apparent that permeability has a marked influence over saturation regardless of the height above the FWL (Figure A.2). Using the J-function it is possible to obtain saturation curves for each porosity and permeability class or for each rock type.

Relative permeability is probably the most challenging input parameter to a simulation model; it is the macroscopic quantity that represents the rock–fluid and fluid–fluid interaction during multiphase flow in the reservoir. The relative permeability reflects the way that the fluid phases are distributed within the porous medium and is governed by the wetting state, pore structure, saturation history, viscosity ratio, interfacial tension, and ratio of the capillary to viscous or gravitational forces.

The *wettability* of the reservoir rock is the most important factor that determines the shape of the relative permeability curve. This can be

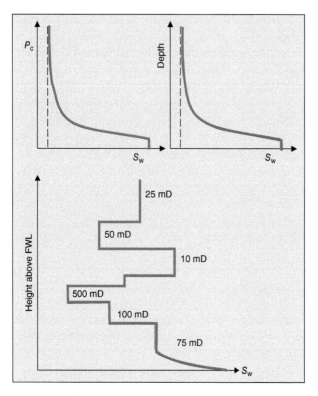

Figure A.2 The relationship between capillary pressure, height, and permeability demonstrating the impact of rock quality on water saturation. *Source:* Cannon. S.J.C. (2016) Petrophysics: a practical guide. John Wiley & Sons, Ltd. © 2016, John Wiley & Sons.

shown in a plot of two-phase oil–water permeability curves for an oil-wet and a water-wet reservoir: the two most extreme examples (Figure A.3). These extremes are seldom seen in practice, but the most striking difference between the curves is the location of the end points. If water is the wetting phase then, at low water saturations, there is little effect on the oil relative permeability because the water remains in all the smaller pores and oil is able to flow easily. Where oil is at irreducible saturation, it forms small disconnected droplets in the larger pores that effectively block the flow of water resulting in a marked reduction in water permeability. Typically, water end point–relative permeability is between

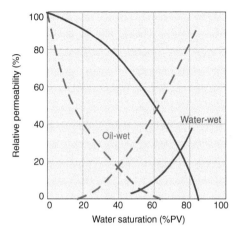

Figure A.3 Typical relative permeability curves for saturation data of oil-wet and water-wet rock types.

0.05 and 0.3 at residual oil saturation of around 25%. If oil is the wetting phase the effects are reversed producing an almost mirror image of the curves. It should be noted that for gas, which is always the non-wetting phase, wettability is less relevant.

Pore structure is a key factor in determining the shape of a relative permeability curve: a poorly sorted, water-wet, consolidated sandstone has a very different profile to a well-sorted, water-wet, unconsolidated sandstone; the difference between a fluvial sandstone and the deposits of a turbidity current (Figure A.4). The main difference in the curves is the range of movable saturations: the consolidated sandstone has much higher end point saturations than the unconsolidated sand; however, this could reflect the difficulty in measurements on unconsolidated material. Typically, residual oil saturation ranges between 25 and 40% and connate water between 15 and 25%; for an unconsolidated sand these ranges could be 10–15% lower. The pore throat ratio and the connectivity of the pore structure are the main controlling factors in residual saturations: the pore throat size controls the 'snapping off' of the non-wetting phase resulting in trapping of the oil in poorly connected pore networks. Where the pores are well connected there are other routes for the oil to migrate reducing the volume of trapped oil and lowering the residual saturation. Relative permeability also depends on the *saturation*

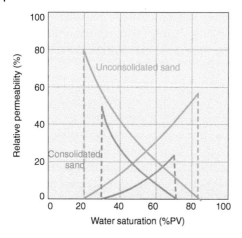

Figure A.4 Typical relative permeability curves for saturation data of consolidated and unconsolidated sandstones.

history of the reservoir and may show the effects of *hysteresis*. However, as a reservoir simulator usually only requires the primary filling cycle, this distinction is not necessary.

Relative permeability is measured either in a steady-state or unsteady-state experiment. In a steady-state experiment, a fixed ratio of liquids is flowed through the sample until pressure and saturation equilibrium is reached; achieving steady-state flow can be time-consuming, especially in less-permeable material. The effective permeability of each liquid is calculated as a function of the relative saturation using Darcy's law, by measuring the flow rate, pressure differential, and saturation. Monitoring the total effluent from a core sample during an imposed flood and calculating the relative permeability ratio that is consistent with that outcome is the basis of unsteady-state measurements. Steady-state experiments are more reliable and accurate but take longer than the cheaper unsteady-state tests, which provide a greater interpretational challenge. Unsteady-state measurements are made using a displacement method whereby a core sample is flooded with 'oil', then flooded with water at a constant injection rate, or gas at a constant pressure differential. The relative permeability of the two phases is calculated from the velocity or pressure difference recorded across the core or from the injection rate. Relative permeability is then calculated from the Darcy equation for single-phase flow together with several dimensionless variables used to infer the impact of different flow parameters such as mobility and movable fluids.

Use of this method requires very accurate measurement of individual phases expressed as fractions of the movable pore volume, flow velocity, or pressure differential, none of which are easy to measure where absolute permeability or mobility is low. There is a third experimental way of measuring relative permeability using a centrifuge: this is an accelerated gravity drainage process suitable for both drainage (gas–oil systems) and imbibition (some water–oil systems). As with the displacement method, the relative permeability is only taken from the end point of the curve. These unsteady-state methods require a complex interpretational process and are subject to much error. Often, all three types of measurement will be available, and a way must be found to integrate the results in a meaningful way: the 'art of reservoir engineering'.

Increasingly, with the development of compositional simulators it is necessary to generate three-phase relative permeability data; this is often done by integrating the results of two-phase measurements of oil–water, oil–gas, and gas–water relative permeability. This is not the place for an in-depth discussion of the mathematical solutions that have been developed.

PVT data are another key input for most simulation studies. Black oil simulators treat the fluids flowing through the pores, oil, gas, and water, as three separate components: light hydrocarbon (l), heavy hydrocarbon (h), and a water element (H_2O). For most black oil simulators (not compositional) the H_2O component contributes to the water phase, while the two hydrocarbon components contribute to the oil and gas phase. Any PVT model in a simulator represents the relationship between the reservoir phases (oil, gas, and water) and the component (l, h, H_2O) properties that are applied to each grid block. Assuming the grid block contains a volume of each hydrocarbon component (x_l and x_h) and of a property (d_l and d_h) then the simulator will make the following calculations.

$$V_o = B_o \frac{x_h - r_s x_l}{1 - r_s R_s}$$

$$V_g = B_g \frac{x_1 - R_s x_h}{1 - r_s R_s}$$

$$\rho_o = \frac{d_h + R_s d_1}{B_o}$$

$$\rho_g = \frac{d_1 + r_s d_h}{B_g}$$

In these equations the volume of oil (V_o) and gas (V_g) and the density of oil (ρ_o) and gas (ρ_g) are calculated from the PVT parameters for each phase (B_o, B_g, R_s, and r_s). From this we get the following relationship:

$$V_o \rho_o + V_g \rho_g = x_h d_h + x_1 d_1$$

If the left-hand side of the equation represents the total mass of the hydrocarbons in the grid block then the right-hand side should represent the same, where the associated mass is the same as the combined light and heavy hydrocarbon component. In this way, the water component does not need to be considered for the purposes of the conservation of mass.

Sampling of reservoir fluids for PVT analysis plays a key role in their applicability. Before taking a sample either during a well test or downhole formation tester it is important to condition the well to maximize the representativeness of the sample. Several samples should be taken in case of contamination or mishandling at the well site. Once a sample reaches the laboratory it is too late to make any corrections and the sample is more than useless as it might be analyzed, and the results used in the simulation even though they are corrupt. Given sufficient experimental data the black oil/volatile oil PVT properties B_o, R_s, B_g, and r_s can be calculated without resorting to a computational simulation; however, some simplifications can be made when data is limited: these include making the assumption that r_s is zero and using material balance, using correlations and material balance, or using a phase behavior package that matches the available data as well as possible. All calculation methods of PVT parameters, apart from the phase behavior solution assume that the surface volumes of light and heavy components are conserved; this is not strictly true but is the assumption made by the simulator.

The determination of the PVT parameters is based on the assumption that there is a relationship between the overall composition and pressure. The standard PVT experiments use this relationship:

1) Constant mass depletion (flash)
2) Constant volume depletion
3) Differential liberation

All three experiments can be described as a series of isothermal volume expansion steps with an associated pressure decline, with each step followed by removing a known volume of gas from the system at constant temperature. Each experiment has the capacity to provide a certain amount of data on the volume of gas (V_g) and oil (V_o) at a given pressure. Once the reservoir fluids have been successfully characterized, they are assigned to individual grid blocks by a series of look-up tables as a function of pressure. It may be necessary to have distinct fluid regimes in a field; in which case it is important that each volume is in thermodynamic equilibrium at initialization.

Initial reservoir pressure data is measured by downhole gauges or open hole formation testers and it is essential to recognize any areas with different pressure regimes. After production commences it is possible to use these tools to determine differential pressure depletion. Matching the grid block pressure with that measured in the field is a key history matching parameter and also essential for the correct operation of the simulator.

Production data are required for history matching a simulation model: oil production rates together with gas and water injection rates. Gas and oil production (GOR and water-cut) are the output of the model calculated from the simulated pressures and saturations. The input data comes from monthly production reports: the formats are usually the same in all regions, only the language changes. Most simulators have preprocessors that allow the monthly rate data to be smoothed or manipulated during periods of erratic production such as start-up or shut down of production. Where fields have a long production history it may be necessary to simplify the history to six-monthly or annual rates of production. Often, production history

data is designed for fiscal accounting rather than engineering purposes making their use less suitable, but still usable with a degree of
caution, especially where gas production (e.g. flaring) is concerned.
However, total field production, or that recorded at a gathering station, is probably more reliable than individual well rates, especially in
larger fields. The reason for this is that individual well allocations are
probably only recorded intermittently when a well test is carried out.
Gas and water injection rates and volumes are usually to be
relied upon.

Well completions are another typical input required for a simulation
study and must be stipulated for each well in the model in order to
allocate total well production to each active grid block. This is essential
for the correct calculation of pressure and saturation in the model.
Mistakes in recording completion intervals and their position in the grid
are a common source of inaccuracy in the results. This factor is more
apparent after well interventions are performed and the model not
correctly updated.

Before the model can be *initialized* a further set of data is required to
establish the equilibrium conditions with respect to pressure and
saturation. In the typical case of an oil reservoir with a primary gas cap
and an aquifer, the following parameters are required:

- Oil reference pressure at a given depth (datum).
- Oil–water contact (OWC) depth. This should be the FWL to correspond
  with the saturation pressure distribution described by the drainage
  water–oil capillary function and the oil and water density in each
  grid cell.
- Gas–oil contact (GOC). Similarly, the GOC is used in conjunction with
  the drainage gas–oil capillary function and gas density to calculate the
  initial gas pressure and saturation in each cell.

In the case of tilted contacts or variable PVT condition, special measures will need to be applied for initialization of the model. Initialization
allows the calculated hydrocarbons in-place to be compared with those
calculated using volumetric methods including a geocellular model. Do
not be surprised by a difference in the results and be prepared to accept
up to 5% variation: more than this suggests a more fundamental error in
one or the other. One way to mitigate errors is to initialize the model in a
stepwise fashion starting with the geological grids and calculate the gross

rock volume (GRV) and compare with the simulation model input. Do this with each of the static inputs including pore volume, NTG and saturation functions, comparing each with the simulation model. From this, it is possible to determine where the discrepancy occurs and to deal with it. Any variation in volume of around 30% will be an error in the top structure map often due to depth correction.

A.2.1 Simulation Grid Construction

This topic has been covered in some detail in the main book; however, the following are worth repeating as building the grid is always a compromise between detail and computer speed or run time of the model. For relatively simple models, prototypes, or sector models, a midpoint or Cartesian grid is recommended, but most modern simulators will accept a corner point grid system that allows curvilinear faults and more complicated geological constructs (Figure A.5). Orientation of the grid should be along the main flow directions. This may be tested with a streamline simulation or with reference to the main geological structures: faults, channels, or fractures. I have found that by starting with a rectangular grid, a more acceptable model is generated: by acceptable, I mean visually not necessarily numerically!

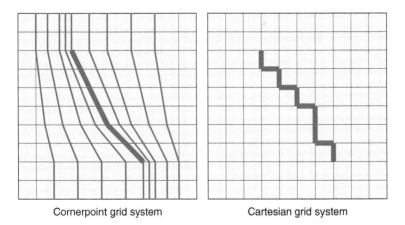

Cornerpoint grid system   Cartesian grid system

Figure A.5   Examples of Cartesian and corner point grid models incorporating a fault.

A *corner point grid* is a logical rectangular grid where the coordinates of its eight corners define each grid cell, but where the cell corners in a column of grid cells are restricted, it may be located on four common grid pillars. The grid pillars are the vertical grid lines connecting a column of grid cells. In a corner point grid, the grid pillars must be straight lines. A corner point grid is defined by specifying two points for each grid pillar, and then by defining the depth values associated with the corners. Each cell is identified by integer coordinates (i, j, k) where the k coordinate runs along pillars, and i and j span each layer. Corner point grids are the norm for all major reservoir simulators; unstructured or PEBI grids have been used for special cases.

A *Cartesian grid* is based on a coordinate system that specifies each point uniquely in a plane by a pair of numerical coordinates that are assigned distances to a point from two fixed perpendicular lines. This grid coordinate system allows the construction of curved surfaces, faults, in the geological grid. A *logical rectangular grid* is defined as a grid where the grid cells can be organized in a (x, y) uniformly regular coordinate system. No faults may be incorporated, and grid pillars are vertical.

As all the grid pillars associated with a column of grid cells must be straight lines, a corner point grid imposes restrictions on what type of faults can be modeled in the grid. Listric faults and Y-shaped faults cannot be modeled with a corner point grid. Note that four corners define a cell surface, and the cell surface is generally not a plane. For calculations involving cell surfaces, this can introduce ambiguity in the location of the cell surface. This is, for instance, the case when calculating the length of a well inside a grid cell where the intersection points between the grid cell and the well must be determined.

Simulation grid design is an optimization process where various aspects must be balanced against each other:

- The grid should be constructed to represent the main geological features as accurately as possible (Figure A.6). This should include representation of:
  - Structural elements as reservoir boundaries, layers, faults, etc.
  - Geological heterogeneities defined at various length scales, such as permeability contrasts, barriers, etc.
- The grid should represent fluid distribution as accurately as possible, including fluid contacts and transition zones.

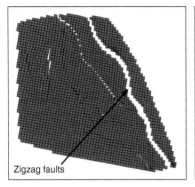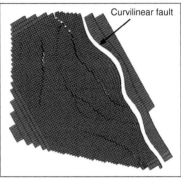

Figure A.6 Grid resolution aligned to the major structural features creates a more robust and realistic faulted grid. *Source:* Cannon. S.J.C. (2018) Reservoir Modelling: a practical guide. John Wiley & Sons, Ltd. Courtesy Emerson-Roxar.

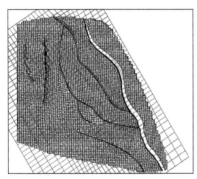

Figure A.7 Example of grid alignment when both geocellular and simulation grid are designed together such that the fine grid fits the coarse grid with minimal upscaling.

- The grid should represent flow geometry as accurately as possible (Figure A.7), using smaller grid cells in areas where important saturation changes may occur.
- The grid should be designed to capture well geometry as closely as possible.
- The grid should be constructed to minimize numerical errors related to discretization of the model equations and thus minimize computation time.

It is not really possible to satisfy all these requirements as they, to some extent, are mutually conflicting.

Selection of the *grid block size* has to be addressed in every simulation study and should be agreed with geologists early in the process; this too is a compromise between run time and geological detail. Grid blocks should not be larger than the characteristic length of the primary geological component; ideally, a grid block should be half the size of the channel width or shoal width where these represent the main flow units. The distance between wells will also control grid size as there should be at least one cell between each well; wells should not be located in adjacent cells and injectors and producers should be separated by at least three cells. Well completions should not be placed within three cells of the GOC or OWC; modeling a horizontal well close to the OWC requires thinner cells to be used between the well bore and the contact. These rules alone can affect the lateral and vertical dimensions of cells in the hydrocarbon-bearing part of the reservoir; it may be possible to increase the size of grid cells in the aquifer. In the 1980s, simulation models with 10 000 cells would be run overnight using the state-of-the-art VAX-Cray computers available. Today, using a desktop PC or even a laptop, a well-built 50 000 cell simulation model will converge on a solution in minutes (maybe up to 60 minutes!).

Numerical dispersion in model is an artifact of the mathematical techniques used in simulation due to rapid changes in saturation and introduces a bias in the results related to cell size and orientation. When there are fewer, larger cells in the model the shape of the saturation profile is smeared out reducing the displacement efficiency of the model at breakthrough. This effect is more pronounced with high mobility fluids. In the case where two equally-spaced injectors are supporting a producer, the orientation of the grid should be normalized so that the distance traveled by the injected fluid is the same, otherwise breakthrough will occur at different times. This is particularly true of high mobility displacement systems such as gas or steam flooding. It is possible to mitigate this effect by using a nine-point formulation of the flow equations so that diagonal vectors are included (Figure A.8).

Grid orientation effects is a collective term used to denote errors introduced in the numerical solution caused by the orientation of the local coordinate axes in the grid. The finite discretization used to approximate the basic model equations introduces a bias in the numerical flow

Figure A.8 Comparison of four-point and nine-point formulations of the flow equations so that diagonal vectors are included.

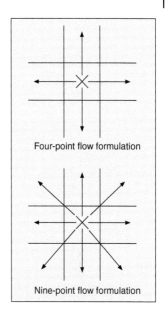

Four-point flow formulation

Nine-point flow formulation

depending on the coordinate directions. Flow in the direction of the grid lines tends to be overpredicted, whereas flow moving diagonally in the grid tends to be underpredicted (Figure A.9).

Grid orientation effects can arise from several sources:

- The nonlinear coefficients in multiphase equations. Grid orientation effects will vanish when the grid cell size approaches zero. The problem increases for increasing mobility ratio.
- Use of a five-point discretization model for non-orthogonal grid cells. The grid orientation effects will not vanish when the cell size approaches zero. The problem can be reduced by application of a discretization molecule with more than five cells involved. Use on nonstructured grids also has a positive effect on the grid orientation problem.

The grid orientation effect will also be present in the vertical direction, where, for instance, the grid will slow down flow along the top horizon in a zone using a base-conformable grid. A proportional grid should be expected to represent the flow more correctly.

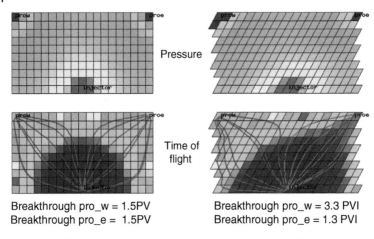

Breakthrough pro_w = 1.5PV   Breakthrough pro_w = 3.3 PVI
Breakthrough pro_e = 1.5PV   Breakthrough pro_e = 1.3 PVI

Figure A.9   Numerical dispersion is created when the simulation grid requires rotation so that the flow paths between wells become distorted. *Source:* Long et al (2002). ©Petroleum Exploration Society of Australia.

The term 'grid orientation effects' is usually reserved for problems found in models with isotropic permeability, but an additional problem arises if the permeability field is anisotropic and the grid orientation does not correspond to the principal permeability axes. Typically, this could arise if the grid is constructed parallel to a well trajectory and not parallel to the geological bedding plane.

There are a few points to remember when building a simulation grid because this should be a key stage of integration when the geologist, petrophysicist, and reservoir engineer should interact constructively. It is important to capture as much of the geological description as is possible but consider that even in a geological model of 500 000–5 000 000 cells each one will probably be no less than 25 m × 25 m and 0.5–1 m thick; already, this is significantly greater than the pore-scale movement of fluids in the reservoir and of the input data. It is important to capture the main barriers to flow, such as faults, and the boundaries of sequences or lateral changes in facies, but once again this is where compromise is king. Fortunately, if the objective of the model is well defined it is possible to build a grid with the correct orientation, dimensions, and number of cells to resolve the problem. Sometimes a well-built grid with more cells will be of more use than a badly designed grid with fewer cells: but this does not always mean that an orthogonal grid is best.

An aquifer is defined as a geological formation that is storing water and also capable of transporting a significant amount of water. The term *aquitard* is used for a formation that may contain water but is not capable of transmitting significant quantities of fluid under normal pressure gradients. If an aquifer is to be modeled as a numerical model, one or more inactive grid cells must be selected to represent the aquifer. The following guidelines should be followed when defining the aquifer:

- If the aquifer volume is large, consider representing the numerical aquifer by more than one grid cell areally.
- Select grid cells with a nonzero bulk volume and a 'reasonable' depth value. This is to avoid artificial effects introduced in the aquifer model.
- To avoid unwanted flow connections, each aquifer should be surrounded by inactive grid cells.
- Avoid connecting a water aquifer directly into a hydrocarbon zone as this can cause instabilities in the model.
- Define separate fluid in-place region numbers (FIPNUM) for the aquifer cells, to be able to supervise pressure and volumes in the aquifer.

A.2.2 Grid Design Workflow

The following steps are considered as basic elements in the workflow:

- Define area covered by the simulation grid, i.e. polygon defining the model boundary is preferably rectangular.
- Select simulation grid fault model: which faults should be included in the model?
- Use of sloping/vertical faults.
- Use of non-zigzag/zigzag faults.
- Define layering from geological model.
- Define areal design grid, including boundary lines. This must be done in an iterative loop with the layering to minimize the total number of (active) cells. Avoid grid editing as this prevents quick updating of the model.
- The (i, j, k) origin should be defined in the 'northwestern' corner of the model, to create a right-handed coordinate system.
- Remove non-orthogonal and thin or twisted grid cells.
- Define fault transmissibility, if required, to represent sub-seismic faults.

- Select cells to be used for numerical aquifers. Large aquifers should be represented by more than one grid cell.
- If necessary, create local grids based on the well trajectories; local grid refinement.

A.2.3   Quality Control of Grids

There will always be imperfect cells in an upscaled grid, but it is the number of these cells and the volume they represent that decides whether they can be ignored, or the grid rebuilt. Perform a visual inspection of the grid, both in 3D and cross sections.

- Check cross sections with correspondence between horizons and grid cells, especially near faults.
- Check areal fault pattern, i.e. midlines from geo-model versus fault splits in the simulation model.
- It should be possible to check simulation grid fault modeling versus the seismic cube in-depth.
- Check distribution of cell thickness. Avoid large numbers of very thin grid cells.
- Check for twisted grid cells ('Inside/out cells'); these must be removed.
- Check for layers with overlapping depth; these must be removed.
- Check for isolated grid cells; these can be removed.
- Check the distribution of minimum and maximum angles for the horizontal projection of top and base horizons. Ideally, this should be centered around 90°.
- Check for non-convex grid cells and cells with a large dip toward faults. Both types of cells are accepted by the reservoir simulator, but should be avoided.
- When cell properties are defined, perform a simulation with no wells in the model to verify that the grid yields a stable model. This is especially important when local grids are included to select between solution methods for these grids.
- If a numerical aquifer model is used, check initial stability of the aquifers.

Always perform a simulation without wells to check stability!

A.3   Upscaling

It is seldom possible for a fine-scaled geological model to be used directly as input for reservoir simulation: the process of upscaling must be undertaken. Upscaling is often an underestimated phase of the workflow; however, by early discussion with all the interested professionals, the process can be made easier and the results more acceptable. By orienting the grid with due consideration of the main flow direction before starting facies modeling, there is less subsequent manipulation of the model. Discussing what size of model is needed for speedy simulation can help to define the scale of the modeling grid. It is essential that upscaling preserves the architecture and pore volume of the net reservoir to ensure that connectivity in the geological model is consistent with recovery estimates predicted by the simulation model. In general, an upscaled grid will have fewer larger cells, coarser vertical layering, and zigzag faults. 'A cynic might describe upscaling as the process of putting incorrect information into the wrong model to get the right answer' (Mike King, personal comm.): just like history matching, it is a nonunique process.

Property upscaling is the process of finding 'effective' values defined on a coarse scale that can represent the distribution of a parameter defined on a finer scale (Figure A.10). The upscaling process must reflect the planned use of the parameter at the coarse scale. For instance, should upscaling of permeability depend on whether the coarse scale permeability is to be used for cell-to-cell flow, cell-to-well flow, or as the basis for establishing a capillary pressure relationship on the coarse scale?

There are four equally important factors in upscaling:

- How heterogeneous is the upscaling region?
- How well are the length scales separated?
- The sampling method used.
- The upscaling technique chosen.

A large number of upscaling methods for flow models exist, and selection of the best method can be difficult. For permeability, we often distinguish between static and dynamic methods:

- Static methods are based on performing arithmetic operations on the property data and the calculation of some sets of statistical averages.

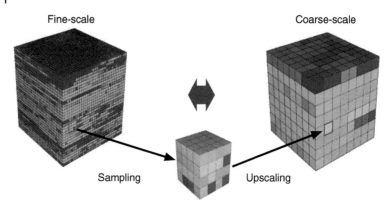

Fine-scale   Coarse-scale

Sampling   Upscaling

"A cynic would describe upscaling as putting incorrect information into a model to the get right answer"   (*Mike King, BP*)

Figure A.10   Upscaling of reservoir properties is dependent on the sampling method selected, scale, and region. *Source:* Cannon. S.J.C. (2018) Reservoir Modelling: a practical guide. John Wiley & Sons, Ltd. Courtesy of Schlumberger-NexT.

- Dynamic methods are based on solving one or several flow equations. The term flow-based upscaling is also used as an alternative.

A.3.1 Statistical Averages

A static method is based on using one of the many statistical averages available (Figure A.11). The most basic methods applied are described below.

- The arithmetic average is the natural average for all volumetric properties such as porosity and shale volume (V_{sh}). It is also the exact value for the horizontal permeability in a perfectly layered reservoir with constant permeability in each layer. A weighted form is often used where the weights can be bulk volume, layer thickness, etc.
- The harmonic average is the exact upscaled value for the vertical permeability in a perfectly layered reservoir with constant permeability in each layer. Note that the harmonic average is zero if a zero value occurs among the samples. The harmonic average can be weighted in the same manner as the arithmetic average.

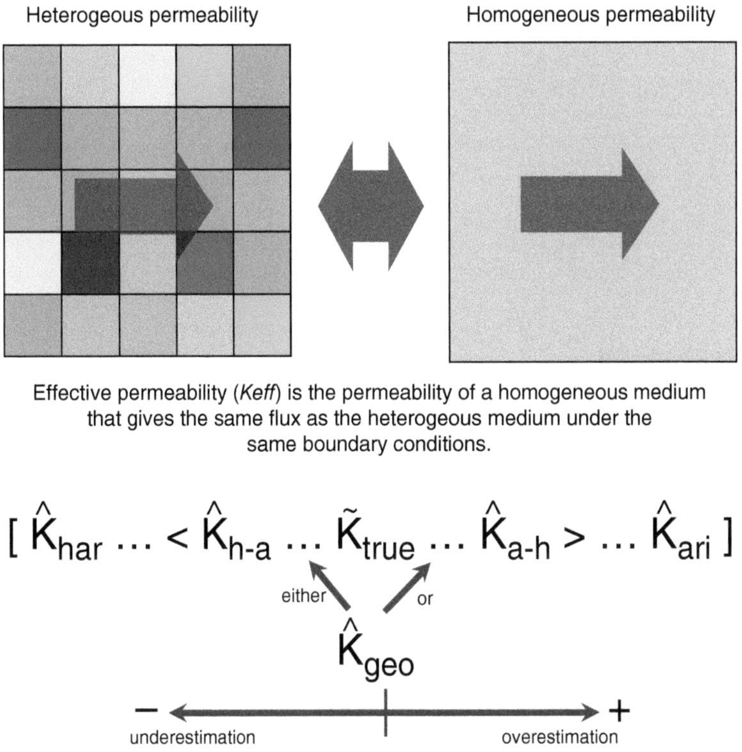

Heterogeous permeability   Homogeneous permeability

Effective permeability (*Keff*) is the permeability of a homogeneous medium
that gives the same flux as the heterogeous medium under the
same boundary conditions.

$$[\ \hat{K}_{har} \cdots < \hat{K}_{h\text{-}a} \cdots \tilde{K}_{true} \cdots \hat{K}_{a\text{-}h} > \cdots \hat{K}_{ari}\ ]$$

either $\nwarrow$ $\nearrow$ or

$$\hat{K}_{geo}$$

$-\ \longleftarrow \quad \longrightarrow\ +$

underestimation   overestimation

Figure A.11   Permeability upscaling is a more challenging task and may
require different methods or dynamic pressure solver techniques. *Source:*
Cannon. S.J.C. (2018) Reservoir Modelling: a practical guide. John Wiley & Sons,
Ltd. Courtesy Emerson-Roxar.

- The geometric average is the exact limit for a region where the permeability is lognormal distributed, and uncorrelated, i.e. there is no coupling between neighboring points in the reservoir and the distribution is totally random.
- The power law average is a generalized average that includes the basic arithmetic, harmonic, and geometric average. Based on the result of any type of upscaling, the corresponding power average value for p can be calculated for comparison with the basic statistical estimates.
- The arithmetic-harmonic average is a combination of arithmetic and harmonic averages (Figure A.12a). First, arithmetic averages are

(a)   (b)

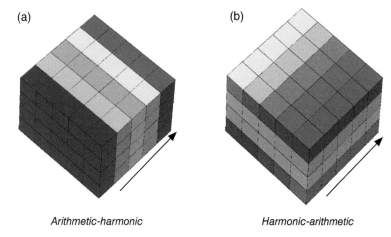

Arithmetic-harmonic   Harmonic-arithmetic

Figure A.12   Examples of two common averaging methods: (a) arithmetic-harmonic and (b) harmonic-arithmetic. *Source:* Cannon. S.J.C. (2018) Reservoir Modelling: a practical guide. John Wiley & Sons, Ltd. Courtesy Emerson-Roxar.

calculated for planes perpendicular to the direction where the permeability is to be calculated. Then, the harmonic average is taken for the arithmetic averages. The arithmetic-harmonic average is the exact upscaled permeability in a perfectly layered reservoir with constant permeability in each layer. That means that it gives the arithmetic average for the horizontal permeability and the harmonic average for the vertical permeability.

- The harmonic-arithmetic average is another combination of arithmetic and harmonic averages (Figure A.12b). First, harmonic averages are calculated along stacks of cells in the direction where the permeability is to be calculated. Then, the arithmetic average is taken for the harmonic averages. For a perfectly layered reservoir with constant permeability in each layer, the harmonic-arithmetic average equals the arithmetic-harmonic average.

Renormalization is an upscaling technique where permeability is modeled by an equivalent resistor network, and Kirchhoff's law for resistor networks is used as an analog to Darcy's law for incompressible fluids. The simulation grid cells are divided into a sub-grid where the dimension in each direction is $2n$, where n is a user-defined parameter. A sub-grid consisting of $2 \times 2 \times 2$ cells is selected, and a pressure gradient is applied in the direction where the permeability is to be calculated. This gives an intermediate upscaled permeability for the sub-grid. Grouping on higher levels continues the process until the upscaled value for the simulation grid cell is achieved.

A.3.2 Dynamic Upscaling

In dynamic upscaling, a flow simulation is performed on the fine-scale grid cells in the upscaling region and it is necessary to specify boundary conditions at the boundary of the region. The result of the upscaling process will generally depend heavily on the boundary conditions selected (Figure A.13).

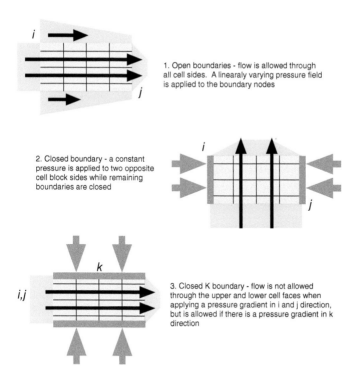

1. Open boundaries - flow is allowed through all cell sides. A linearaly varying pressure field is applied to the boundary nodes

2. Closed boundary - a constant pressure is applied to two opposite cell block sides while remaining boundaries are closed

3. Closed K boundary - flow is not allowed through the upper and lower cell faces when applying a pressure gradient in i and j direction, but is allowed if there is a pressure gradient in k direction

Figure A.13 Different boundary conditions that can be applied to pressure solver upscaling of permeability. *Source:* Cannon. S.J.C. (2018) Reservoir Modelling: a practical guide. John Wiley & Sons, Ltd. Courtesy of Schlumberger-NexT.

The boundary problem is one of the fundamental problems of upscaling. Ideally, the boundary conditions applied in the upscaling should reflect the boundary conditions the coarse cell experiences as part of the total grid during the full-field simulation. For a complex field simulation these boundary conditions will vary both in time and space and cannot be properly anticipated.

For *no-flow boundary conditions* a pressure gradient is applied along each of the coordinate directions, one at a time. The following conditions are assumed:

- Constant pressure over the surfaces perpendicular to the pressure gradient.
- No flux through the surfaces parallel to the pressure gradient.
- This is an analog with a laboratory core-flood experiment.

For *linear boundary conditions*, the pressure varies on the boundary in a linear fashion. Here, the pressure applied is a pressure gradient applied in a fixed position lying inside the simulation grid cell.

For *periodic boundary conditions*, it is assumed that the fine-scale permeability is identical in all the simulation grid cells and that the flow is not influenced by the wells. A pressure gradient is applied along each of the coordinate directions one at a time. The following boundary conditions will then be satisfied:

- The pressure on opposite fine cells differs with a constant on the surfaces perpendicular to the pressure gradient.
- The pressure on opposite fine cells is equal on the surfaces parallel to the pressure gradient.
- The flux through opposite fine cells is equal on all the surfaces.

When assuming that the flux between the fine and coarse scale is conserved for the no-flow and linear boundary conditions and the dissipation (loss of mechanical energy per unit time) for the periodic boundary conditions, an expression of the effective permeability can be derived.

A.3.3 Comparison of Upscaling Methods

Deciding which of the methods to use is often a matter of trial and error, and always dependent on the available data; if you have well test or production data then it is a case of trying to match those results. Where you only have empirical data then the simple option is good enough.

The arithmetic average is the theoretical upper bound for the upscaled permeability, while the harmonic average is the theoretical lower bound. The geometric average is always lower than or equal to the corresponding arithmetic value and higher than or equal to the harmonic value for the same data set. The arithmetic-harmonic average is a finer upper bound than the arithmetic average, while the harmonic-arithmetic average is a finer lower bound than the harmonic average.

Flow-based upscaling using the no-flow boundary conditions will give a lower bound for the effective permeability, while the linear boundary conditions lead to an upper bound. The periodic boundary conditions yield an intermediate value. The linear and periodic boundary conditions yield a full permeability tensor.

The statistical averages are less time-consuming than the flow-based upscaling, but the accuracy is generally better for the flow-based upscaling.

A.3.4   Local, Regional, and Global Upscaling

Upscaling methods can also be classified as local, regional, or global. Local methods are generally less expensive than regional and global methods, but the accuracy is better for regional and global methods.

In using *local methods*, the upscaling is performed cell-by-cell for the coarse grid. Each coarse cell is treated as a more or less isolated region, and only the fine-scale cells that fit inside each coarse cell are taken into consideration in the process. The result usually depends critically on the flow pattern assumed for the coarse cell, i.e. the boundary conditions applied to the fine cell region in the upscaling.

Regional upscaling is similar to local methods, but the influence of the neighboring fine cells is accounted for within a limited region around each flow cell; this is done by creating a 'flow jacket' of fine cells around the focus cell. The regional methods are used to reduce the influence of boundary conditions on the upscaling result.

In *global upscaling* all flow cells are upscaled simultaneously by computing the complete pressure field of the geo-model and performing a 'history match' for the coarse cell properties. The method requires a flow simulation performed on the geo-grid.

A.3.5 Sampling for Upscaling

Sampling is a preprocessing step for the upscaling and is time-consuming and error prone. Selection of the correct sampling method will significantly influence the upscaled result. Reduction of sampling errors is an important goal both for geo-grid design and simulation grid design. The sampling method selected may influence the upscaled result significantly!

The sampling problem consists of finding representative values from the geological grid located inside a cell in the simulation grid. This includes defining the correct volume associated with each fine cell inside the coarse cell. It is especially important to focus on sampling problems for fluvial reservoirs, where channel belts in the geological model may have a width close to the length scale used by the simulation model. For these reservoirs, the geological grid and the simulation grid should be designed with as much correspondence as possible.

The sampling process is rather time-consuming. If possible, several properties should be upscaled simultaneously, using the same sampling process. Two common methods for sampling are commonly defined (Figure A.14):

- Resampling
- Direct sampling

In *resampling*, the simulation grid cell is divided into a finer grid with a specified resolution. Each finer grid cell is assigned the property value that the geological model has at the center of the fine cell. The sampling error will decrease as the sampling resolution increases, but so will the calculation time. It is recommended to have a resolution of no less than four cells for the flow-based upscaling methods.

The simulation grid cell is assigned cells from the geo-model in direct sampling. In most software there are again two different methods:

- Cell center-based
- Cell corner-based

For the cell center-based method the geological grid cells with their center position inside the simulation grid cell are assigned to that cell. For renormalization and the flow-based upscaling methods, the smallest box of geo-model grid cells containing the cell center-based cells is used. Only the part of the geo-model grid cells that is inside is

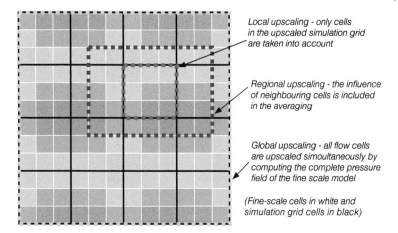

Local upscaling - only cells in the upscaled simulation grid are taken into account

Regional upscaling - the influence of neighbouring cells is included in the averaging

Global upscaling - all flow cells are upscaled simoultaneously by computing the complete pressure field of the fine scale model

(Fine-scale cells in white and simulation grid cells in black)

Figure A.14   Examples of different upscaling regions: local, regional and global. *Source:* Cannon. S.J.C. (2018) Reservoir Modelling: a practical guide. John Wiley & Sons, Ltd. Courtesy of Schlumberger-NexT.

used for cell corner-based. The cell corner-based method can be time-consuming if the number of truncated geo-grid cells is large. The method is only available for some of the statistical averages.

A.3.6   Selection of Sampling Method

Resampling is generally faster than direct sampling, but the sampling error may be significant if there is complex grid geometry or different grid orientations between the simulation model and the geo-model. Generally, it is recommended to use direct sampling. Exceptions from this are:

- When upscaling with renormalization. Direct sampling will fail if the dimension of the sampling grid is not a power of 2.
- When the simulation model and the geo-model are perfectly aligned and have proportional dimension.
- Sampling errors are reduced if a correspondence between sets of layers in the geo-model and a layer in the simulation model can be specified. This is called layered sampling and should be used whenever possible.
- If layered sampling cannot be used, cell corner-based direct sampling should be preferred if possible.

A.3.7   Upscaling Porosity

The porosity is a volumetric property and the purpose of the upscaling is to create a representation of the pore volume distribution found in the geocellular model. Using a simple bulk volume weighted arithmetic upscaling for the porosity does this.

Note that if the geocellular model includes a NTG distribution, the effective porosity should first be calculated and used as basis for upscaling.

A.3.8   Upscaling Permeability

Upscaling of the permeability is especially difficult since the permeability is a nonadditive property. A lot of possible techniques exist, and the results from these techniques can vary significantly. It is generally difficult to decide which method is best for a given model.

When the areal resolution of the geo-grid and the simulation grid is the same, the upscaling is only carried out in the vertical direction. This is often the case for fluvial reservoirs. Then the arithmetic and harmonic average should be used for the horizontal and vertical permeability, respectively.

Simple averaging techniques are less time-consuming than flow-based upscaling, and for many purposes these methods will suffice. This includes, for instance, when upscaling is a part of a model-ranking process. Static techniques can also be sufficient for simpler models, for instance, for map-type models having only a single grid-layer in each zone.

A good choice for a simple averaging method is to use arithmetic-harmonic average for the horizontal permeability and harmonic-arithmetic average for the vertical permeability. For more complex geo-models where it is important to conserve structure from the geo-model into the flow model, flow-based upscaling is recommended.

If two alternative upscaled values are generated, using the no-flow and linear boundary conditions, respectively, lower and upper bounds of the permeability are defined. If the two distributions, in some sense, are close, a representative effective permeability distribution is found, and you can use either of the two. If significant differences between the two generated distributions exist, you will have to decide which element of the reservoir's performance is the most important and choose the boundary condition that preserves these flow characteristics.

The no-flow boundary condition reduces the permeability of a sand and mud mixture, makes shales thicker, and makes channels narrower and more disconnected. The method, is however, very good at including the effect of barriers. If there is a horizontal barrier, there will be no vertical flow in the grid cell. With linear boundary conditions, streamlines can enter and leave every boundary, so there will be a flow around the barriers.

If a barrier exists over a horizontal well and the productivity of the well is to be evaluated, linear boundary conditions are better for assessing the vertical permeability. On the other hand, if the potential for gas coning down to the well is to be evaluated, a vertical flow barrier may be preferable.

A.3.9   Upscaling Net:Gross

There are several reasons why the net:gross (NTG) ratio should not be upscaled as an individual entity:

- NTG is treated both as a volumetric property and as a (horizontal) flow property in the simulations. The upscaling cannot account for both these properties.
- The NTG is used as a multiplier when calculating pore volumes and transmissibility factors. A product of two quantities cannot be averaged individually to achieve the average of the product.

To handle NTG in the upscaling, the quantity should be multiplied into the porosity and the horizontal permeability in the geo-model before the upscaling, thus performing the homogenization on the effective values.

Note that vertical permeability is not affected by the NTG ratio. This requires that the k_v is the true effective value for the cell, not only the net sand value. If it is desirable to have an upscaled porosity that is not affected by the NTG ratio in the model, the following relationships can be used for the NTG ratio and the horizontal permeability:

$$\overline{NTG} = \frac{\overline{\emptyset.NTG}}{\overline{\overline{\emptyset}}}$$

$$\overline{K_H} = \frac{\overline{K_H.NTG}}{\overline{NTG}}$$

A.3.10 Water Saturation Modeling

The water saturation modeling for the simulation model has several objectives:

- Correct volumetric representation.
- Correct representation of end point saturation values for relative permeability and capillary pressure curves.
- Correct representation of mobile and immobile fluids in the transition zone.
- A numerically-stable model without spurious fluxes.

A common experience from many fields is that the water found in the 'transition zone' is not producible. This is not in accordance with a capillary pressure model, where the water should be movable even for reasonable pressure gradients. The water transition zone is probably influenced by upward- and downward-moving fluid levels over geological time, creating a combination of drainage and imbibition effects that cannot easily be explained by a single capillary curve. The capillary pressure curve is basically scale dependent and should ideally be upscaled from the core plug level to reservoir level.

If the J-curves defined by geology yield a reasonable volume representation when used directly in the simulation model, this simplification could be used as a pragmatic approach. J-curves or capillary pressure curves can give the best model for the influence of heterogeneities on tail production. The flow in the transition zone will also be represented more correctly.

A.3.11 Quality Control

When comparing the results remember that the upscaled properties on the simulation model can have been affected by the NTG ratio and/or have been weighted with some volumes.

- Check min/max and histogram distribution between geo-grid and simulation grid for porosity, permeability, and $k_V = k_H$.

- Check pore volume and fluid volumes in defined regions and zones for geo-model and simulation model.

Data from the geocellular model and the simulation model can both be compared with the original data from the well logs. However, in the comparison, remember that the upscaled properties in the wells in the simulation model also include information from the geo-model outside the wells.

Streamline simulation can be used for testing the validity of an upscaling. Streamlines produced using the geo-model can then be compared with streamlines from the upscaled model to see if the main features of the geo-model have been conserved in the upscaling. The streamline simulator can be especially useful for fluvial reservoirs to see if channel communication characteristics are preserved in the upscaling.

The results from the streamline simulator should only be used for comparison purposes and can never replace a traditional numerical simulation for production forecasts. Time estimates given by the streamline simulator have nothing to do with physical time and should again only be regarded as relative estimates. The results from the streamline simulations quickly tend to get messy and difficult to interpret. To be able to analyze the results, it is critical to select a simple artificial well pattern in the simulations. Use a set containing only a few injectors and producers.

A.3.12 Workflow Specification

Sections 12 and 13 specify data and tasks involved when defining property data for the reservoir simulation model. This involves flow properties for simulation grid cells and cell boundaries. As for the simulation grid data, property data should be documented fully:

- Upscaled cell values for porosity and permeability.
- Upscaled cell values for shale volume if that is being used for NTG estimation.
- Upscaled cell values for water saturation.
- Transmissibility multipliers for faults.
- Transmissibility multipliers representing restrictions to vertical flow.
- Formation compressibility.

A.3.13 Upscaling Workflow

The sampling part of upscaling is critical for the final result. The sampling depends heavily on the relationship between the geo-grid and the simulation grid, and this aspect must be taken into account when the two grids are designed.

- Select upscaling methods.
- Use pore volume weighted arithmetic average for porosity.
- For permeability, general rules are not easily given as methods must be based on flow characteristics. For ranking purposes, simple methods as statistical averages should be used. In many cases, a flow-based upscaling is preferable for flow simulations.
- For water saturation, two alternative methods are described where property modeling is discussed. The first method is based on using the J-curves defined as part of the property analysis workflow, and the second method uses upscaling of the water saturation distribution in the geo-model, with a pore volume weighted arithmetic average.

A.3.14 Selected Sampling Method

Direct sampling is usually preferable, although more time-consuming (Figure A.15). If correspondence between geo layers and simulation layers is used, as recommended in the simulation grid design, then layered sampling should be applied. If this is not the case, a volumetric-weighted averaging technique should be used.

- If the endpoint scaling option is used for water saturation, define water saturation table data.
- When the water saturation model has been defined, fluid volumes can be compared with the geocellular model.
- Calculate transmissibility multipliers for faults represented in the simulation grid using a fault multiplier.
- If effects of small sub-seismic faults are to be represented in the model, an updated permeability field should be calculated, accounting for faults not represented in the grid.
- Define transmissibility multipliers for zone boundaries.

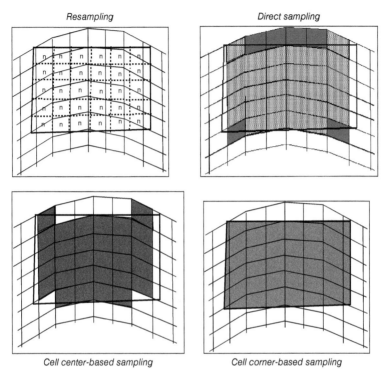

Figure A.15 A comparison of resampling and direct sampling methods for cell center-based and corner-based cases. *Source:* Cannon. S.J.C. (2018) Reservoir Modelling: a practical guide. John Wiley & Sons, Ltd.8). Courtesy Emerson-Roxar.

A.4 History Matching

History matching is essentially the validation of the model by simulating past performance of the reservoir and comparing it with the actual historical data. If differences are found in the simulation, modifications can be made to the input data to try and improve the match, but these should not be made without due reference to the provider of the source data.

History matching is an iterative process and the results are nonunique: it is possible to get the same result by varying different inputs. Both full-field history matching, and a well-by-well basis should be carried out.

History matching can also be looked upon as a way of checking the model's sensitivity to variations in the input data. In this sense, it is a way of improving the understanding of the reservoir description and how the field will perform under different production scenarios. It is worth considering the two classes of history matching variables: the input or control variable and the output and response variable. We change a control variable such as pore volume and see the response in terms of hydrocarbons in-place. When the model has been initialized it is advisable to make several runs through the whole field production history to tune the simulator for efficient running and to gain some idea of the model's sensitivity to some major parameters: it may be wise to establish some sensitivity coefficients against which subsequent changes can be judged.

A history matching exercise usually consists of several steps starting with an overall match in terms of pressures and total production/injection volumes (Figure A.16). During this stage, only overall reservoir parameters should be adjusted, such as aquifer size, relative permeability curves, and PVT parameters. In particular, any changes in the pressure trends should be looked for as they indicate a bubble point. A further stage would be to consider different regions of the field or segments of the model and making regional adjustments to properties or changing the transmissibility of faults or other barriers. Once the response of the model to gross fluid movements and overall performance has been established, individual well performance should be matched. There are many performance matching parameters that can be played with including the well completions, local reservoir properties, and pseudo-relative permeability curves. The use of well lift curves is recommended during this stage as a confidence booster for the next important step which is the prediction runs. Needless to say, it is crucial to keep a log of each run and the changes made; these should be stepwise and individual so that the impact of a change can be monitored. Automatic history matching tools are now available to the industry, but like any form of automation or artificial intelligence (AI) they are only as good as the input data.

The criteria against which to judge a history match depends to some extent on the objectives of the simulation study, but these should be limited to only two or three of the many options and these should be ones in which the measurement uncertainty in the field warrants their inclusion.

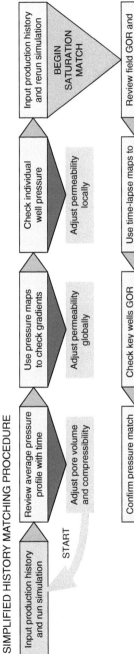

Figure A.16   Simplified history matching workflow from an initial pressure match to a subsequent saturation match.

Pressure data is a key matching criterion but also one with some degree of uncertainty. Historical pressures may be flowing or static measurements and are usually an average for a defined area at a given depth. Formation pressure measurements give the pressure of the mobile phase in the reservoir at specific depths. The pressure reported by the simulator may be a pressure at a given depth or an average pressure over the whole reservoir or phase volume, and only at a few grid points: the midpoints of the cells where the well completions are located. It is then obvious that to compare like with like two corrections are often required: one is a correction to compare the grid block pressure with the calculated flowing pressure in the wells and, second, if using wireline formation test data, a correction for the possible difference between the reference pressure and the flowing phase pressure.

Production ratios (GOR, BSW, WOR, WGR, CGR) are commonly used as matching criteria, but it is essential that these are recorded accurately, which is not always a given as they are determined by the most inaccurate rate measurement. Any measurements that include gas content are especially suspicious. Matches after a phase breakthrough tend to be the most reliable and give confidence in the relative permeability description.

Breakthrough times are notoriously difficult to match and local changes to permeability or transmissibility are required to get a match.

Phase contact movements are difficult to define in a diffuse flow model and are equally challenging to match in a simulation model. Sometimes observed changes in the fluid contacts are used qualitatively rather than quantitatively.

Well productivity matching is often ignored: because the data is used in the history matching phase the rates are imposed on the wells and not freely obtained. For oil wells, the productivity index (PI) should be used as a matching criterion and for gas wells the inflow performance curve. Failure to do this will lead to problems when moving to the prediction part of a study.

Production logging tool (PLT) data are useful for modeling well inflow but not suitable as matching criteria as they are a function of the permeability thickness of the completion rather than the overall performance of the well.

Temperature is a useful matching parameter in thermal simulations only.

Whichever matching criteria are selected there is a natural limitation on the degree of adjustment that can be made to a variable; this depends on the accuracy with which it has been measured in the first place and the upscaling of a parameter. Properties like the PVT are assumed to be measured with such a degree of accuracy that there is little scope for adjustment: the same cannot always be said for the sampling! Another consideration on selecting a matching parameter is the influence it may have on the outcome of the study. Therefore, favorite matching criterion are the shape of the rock curves, or something that has an easily recognized influence such as changing the transmissibility of a barrier (Table A.2).

Of the common history matching parameters used by reservoir engineers, those associated with the reservoir description are commonly the most used, and dare I say abused? Before changing the transmissibility of a fault or barriers always check with the geoscientists that there is indeed something to alter, because a new interpretation may have been undertaken that makes it unnecessary. One field in the North Sea that had been in production for 20 years contained a major fault interpreted on seismic that appeared to separate a line of injectors from the producing wells. The reservoir engineers had to create breaks in this bounding

Table A.2   Typical uncertainty ranges of common history matching parameters.

Typical uncertainty ranges in history matching parameters	Range
Reservoir description:	
NTG ratio	0.7–1.5
Permeability	0.1–10
K_v/K_h ratio	0.1–5
Transmissibility barriers	0.01–100
Depth	0.99–1.01
Reservoir contents:	
Hydrocarbon pore volume	0.7–1.5
Aquifer size	0.01–10
Aquifer permeability	0.1–10
Well description:	
Total permeability thickness product	0.2–5
Permeability thickness distribution over layers	0.4–2.5

fault to enable the match in the simulation. The geophysicist acquired new seismic and a new depth conversion algorithm that moved the boundary fault to the other side of the line of injectors and solved the engineers' problem. Needless to say, the two sets of professionals worked on separate floors of the office and never spoke! It needed an independent advisor to spot the problem when they asked to see the map the engineers were using!

When performing a history match it is worthwhile establishing the expected quality of the outcome; it is called managing expectations. For instance, it may be appropriate to state that an acceptable match is achieved if total water production is within 3% of the recorded amount or that the water-cut trends are matched in 80% of the wells. There is no standard procedure for history matching because each field is unique in geology, reservoir quality, production history, exploitation strategy, etc. therefore each study will require its own solution. However, to begin with, establish the global pressure regime and individual fluid gradients and then match individual well behavior. Where there are many wells in a study it is best to establish a set of key wells: those with the most complete and reliable datasets.

It is at this stage that the engineer is most tempted to adjust the pore volume and permeability both globally and locally. Having established a satisfactory pressure match, move on to the fluid saturation match: check the field-wide GOR and water-cut history and review the relative permeability to improve the match. By studying the movement of fluids and comparing them with model saturations it is possible to recognize any inconsistency in the flood fronts especially around individual wells: this justifies adjustment of individual relative permeability data usually associated with a change in rock properties locally. The model can be considered successfully matched when the main controlling mechanisms of the reservoir have been correctly simulated; individual wells may never be correctly matched. There is a case to be made that given that the simulation is inherently uncertain, the model should be viewed as a probabilistic tool; this is certainly the case when the reservoir description is increasingly the result of stochastic modeling methods.

A.4.1 Prediction Runs

The different scenarios for making predictions of future production should have been established as one the objectives of the study. As the study evolves a few alternatives may suggest themselves but too many would indicate that the purpose of the model was probably poorly defined at the outset. Future production forecasts are made to test different operating strategies and to evaluate the economic impact of these strategies. Prediction runs usually require restarting the model at the point that the history match ends. It is important to remember that the value of the prediction runs is dependent on the data on which the model has been matched.

The number and type of prediction cases depends on the field circumstances and the time and resources available, but it is common to first establish a base case by making a production forecast under the current operating conditions against which other strategies can be tested and evaluated economically. All other scenarios must conform to the technical, operational, and legislative conditions at the time and especially where a large investment might be required. This particularly applies to projects to enhance recovery that might require additional infrastructure or a reduction in product sales due to reinjection of liquid natural gas as part of the water alternating gas (WAG) project, for instance.

Appropriate guidelines and constraints must be put in place before starting prediction runs, these include surface facility limitations and individual well performance. Typical surface constraints are maximum oil, gas and water rates of production, maximum gas and water injection rates and pressure, as well as surface pressure limits and GOR rates. Typical well constraints would include WOR, GOR, total liquid rate, and a minimum oil production rate. Modern reservoir simulators have the necessary flexibility to let the simulation engineer define the working conditions that apply to each prediction run; these can be applied on individual wells, groups of wells, or at the field scale. When comparing several scenarios, it is important to ensure that the same constraints are applied to ensure consistency in their evaluation (Table A.3).

The way in which the simulator works in the history matching stage and the prediction stage is fundamentally different: in the former, field performance is known and the model translates the imposed oil rates into grid block pressures using a well management process. During the prediction stage, rates are unknown and must be calculated by imposing rules on the model usually represented by flowing conditions at the wellhead. Because this boundary pressure must represent different flowing pressures and conditions at the surface, at the completion and in the

Table A.3 Typical constraints for reservoir prediction runs.

Field/Group constraints	Well constraints
Max oil production	Max GOR
Max water production	Max WOR
Max GOR	Max total liquid rate
Max water injection rate	Min and max oil production rate
Max water injection pressure	Min and max water injection rate
Min average reservoir pressure	Min bottomhole pressure
Separator pressure	Well head flowing pressure

reservoir a wellbore hydraulic model is required. This model tries to define the vertical flow performance (VFP) by a series of tables generated to describe the outflow conditions of the wells. Outflow performance has a large impact on well deliverability, so the input parameters must be well thought through, usually with the help of production engineers. Typically, the VFP tables are calibrated against flowing bottomhole and tubing head pressure data from the key wells, where available. Inflow well performance from well test data are equally important and can give actual information on the PI of individual wells. Every effort should be made to use all available flow data from individual wells to characterize well performance in the prediction stage.

The first step in making a prediction run usually involves adjusting the PI of each well in the model to the field data. This is often done by increasing the well skin factor, which is normally set to zero at the start of the modeling study. This is also required to get a 'smooth' transition from the history matching phase as discontinuities are only now seen in individual well rates and pressures. Because the wells operate under imposed rate conditions during the history match, these differences are not seen until this final step in the model process.

A.4.2   Interpretation and Reporting

It is preferable that reporting is a continuous process throughout the project with major stages agreed and approved before moving forward. Normally, regular progress reporting is required by clients and management alike which makes decisions and problems explicit and also reduces reporting time at the end of the study. The main requirement of a simulation study report is that it gives an unambiguous and auditable account of the model's assumptions and construction, and the results and files generated during prediction runs. Details of the geology, petrophysics, PVT data, and relative permeability curves can be included in an appendix. The people who need to be influenced by the study will only ever read the introduction and summary of results.

A.5   Summary

A successful reservoir simulation model is one where the subsurface professionals work together and heed what the facilities engineers say is possible. And never forget 'all models are wrong, some are useful' (Box 1979).

References

Ahuja, B.K., Bahar, A., Kerr, D.R., and Kelkar, M.G. (1994). Integrated Reservoir Description and Flow Performance Evaluation of Self Unit, Glen Pool. SPE/DOE 27748 – Society of Petroleum Engineers.

Amerada Hess (2002). Equatorial Guinea fields: www.hess.com/worldwide/africa/egfields.htm.

Awan, A.R., Teigland, R., and Kleppe, J. (2008). A survey of North Sea enhanced-oil-recovery projects initiated during the years 1975 to 2005. *SPE Reservoir Evaluation & Engineering* 11 (3): 497–512.

Bacon, M., Simm, R., and Redshaw, T. (2003). *3-D Seismic Interpretation*. Cambridge: Cambridge University Press.

Bear, J. (1972). *Dynamics of Fluids in Porous Media*. New York: Elsevier.

Beuthan, H.C., Shibani, M., and Bremeier, M. (2006). Business Improvement by Integrated Concession Development. *SPE 99703* – Society of Petroleum Engineers.

Boe, A., Nordtvedt, J-E., and Schmidt, H. (2000). Integrated Reservoir and Production Management: Solutions and Field Examples. *SPE 65151* - Society of Petroleum Engineers.

Box, G.E.P. (1979). *Robustness in the strategy of scientific model building*. Technical Summary Report ≠1954, University of Wisconsin–Madison.

Brand, P.J., Clyne, P.A., Kirkwood, F.G., and Williams, P.W. (1995). The Forties Field – 20 years young. In: Proceedings of the Offshore Europe Conference, Aberdeen, p. 695–704.

Reservoir Management: A Practical Guide, First Edition. Steve Cannon.
© 2021 John Wiley & Sons Ltd. Published 2021 by John Wiley & Sons Ltd.

Bremeier, M., Elashahab, B.M., and Goh, T. (2005). Sustainable management of a large oil field in Libya. *SPE 93193* - Society of Petroleum Engineers.

Bryant, S.L. and Lockhart, T.P. (2002). Reservoir engineering analysis of microbial enhanced oil recovery. *SPE Reservoir Evaluation & Engineering* 5 (5): 365–374.

Cannon, S.J.C. (1994). Integrated facies description. *DiaLog* 2 (3): 4–5. (reprinted in: *Advances in Petrophysics – 5 Years of DiaLog 1993–1997*, London Petrophysical Society, 1999).

Cannon, S.J.C. (2016). *Petrophysics: A Practical Guide*. Chichester, UK: Wiley.

Cannon, S.J.C. (2018). *Reservoir Modelling: A Practical Guide*. Hoboken, NJ: Wiley.

Carlson, M. (2003). *Practical Reservoir Simulation*. Tulsa, OK: PennWell Corporation.

Carter, A. and Heale, J. (2003). The Forties and Brimmond Fields, Blocks 21/10, 22/6a, UK North Sea. In: *United Kingdom Oil and Gas Fields, Commemorative Millennium Volume* (eds. J.G. Gluyas and H.M. Hichens), Memoir 20, 557–561. London: Geological Society.

Cipolla, C.L., Shucart, J.K., and Lafitte, J.R. (2005). Evolution of frac-pack design, modeling and execution in the Ceiba Field, Equatorial Guinea: SPE Annual Technical Conference and Exhibition, Dallas, SPE 95514, p. 15.

Claiborne Jnr., E.B., Malone, B.P., and Marshall, J.C. (2002). Ceiba completion optimization: A fast track approach to success. *SPE 77434* – Society of Petroleum Engineers.

Clampitt, R.L. and Reid, T.B. (1975). An economic polymerflood in North Bank Unit, Osage County, Oklahoma. *SPE 5552* – Society of Petroleum Engineers-AIME

Clayton, F.M., and Gordon, N.C. (1990). The Leman F and G development: Obtaining commercial production rates from a tight gas reservoir. In: *Proceedings Europec 90 Conference, The Hague*. SPE-20993, p. 429–440.

Dailly, P., Lowry, P., Goh, K., and Monson, G. (2002). Exploration and development of Ceoba Field, Rio Mini Basin, Southern Equatorial Guinea. *The Leading Edge* 21: 1140–1147. Society of Exploration Geophysicists.

Dake, L.P. (2001). *The Practice of Reservoir Engineering*. Burlington: Elsevier BV.

Droegemueller, U. and Leonhardt, B. (2005). Hydraulic frac stimulations in a Libyan oil field – A case history. *SPE 95019* – Society of Petroleum Engineers.

Dyck, W.C. and Moore, R.M. (1994). Low-pressure gas field operations in Leman. In: *Proceedings SPE European Petroleum Conference, London*. SPE 28889, p. 233–244.

Ezekwe, N. (2012). Advances in reservoir management technologies. SPE Distinguished Lecturer Program

Freeman, P., Kelly, S., MacDonald, C. et al. (2008). The Schiehallion Field: Lessons learned modelling a complex deepwater turbidite. In: *The Future of Geological Modelling in Hydrocarbon Development*, vol. 309 (eds. A. Robinson, P. Griffiths, S. Price, et al.) Geological Society Special Publication, 205–219. London: Geological Society.

Frorup, M., Jenkins, C., McGuckin, J. et al. (2002). Capturing and preserving sandbody connectivity for reservoir simulation: Insights from studies in the Dacion Field, Eastern Venezuela. *SPE 77593* – Society of Petroleum Engineers.

Helgeson, D., Towart, J., Rose, P, and Gibson, J. (2019). Extending the Beryl Field Life: Recent exploration success at Callater, Storr, Corona, & Garten. Abstract – PESGB Evening Lecture (Aberdeen)

Hillier, A.P. (1990). Leman Field. In: *Structural Traps I* (eds. E.A. Beaumont and N.H. Foster), 51–71. Tulsa, OK: Treatise of Petroleum Geology: American Association of Petroleum Geologists.

Hillier, A.P. (2003). The Leman Field, Blocks 49/26, 49/27, 49/28, 53/1, 53/2, UK North Sea. In: *United Kingdom Oil and Gas Fields, Commemorative Millennium Volume* (eds. J.G. Gluyas and H.M. Hichens), Memoir 20, 761–770. London: Geological Society.

Hillier, A.P. and Williams, B.P.J. (1991). The Leman Field, Blocks 49/26, 49/27, 49/28, 53/1, 53/2, UK North Sea. In: *United Kingdom Oil and Gas Fields, 25 Years Commemorative Volume* (ed. I.L. Abbotts), Memoir 14, 451–458. London: Geological Society.

Horseman, C., Ross, A., and Cannon, S. (2014). The discovery and appraisal of Glenlivet: a West of Shetlands success story. In: *Hydrocarbon Exploration to Exploitation West of Shetlands*, vol. 397 (eds. S.J.C. Cannon and D. Ellis) Geological Society Special Publication, 131–144. London: Geological Society.

Johnson, C.L. (1992). Burbank Field-USA, Anardarko Basin, Oklahoma. In: *Stratigraphic Traps III (Treatise of Petroleum Geology Atlas of Oil and Gas Fields)*, vol. 1992 (ed. N. Foster), 333–345. Tulsa, OK: American Association of Petroleum Geologists.

Jones, R.D., Rose, J., Lurie, P. et al. (1997). Design, planning and implementation & management of a multi-lateral well on the BP Forties Field – a North Sea case history: In: *Proceedings Offshore Europe Conference, Aberdeen*, SPE 38494, p. 227–241.

Kelkar, M. and Richmond, D. (1996). Implementation of reservoir management plan – Self Unit, Glenn Pool Field. *SPE/DOE 35407* – Society of Petroleum Engineers.

Kerr, D., Ye, L., Bahar, A. et al. (1999a). Glenn Pool Field, Oklahoma: A case of improved production from a mature reservoir. *E & P Notes AAPG Bulletin* 83 (1): 1–18. American Association of Petroleum Geologists.

Kerr, D., Ye, L., Aviantara, A., and Martinez, G. (1999b). Application of borehole imaging for reconstruction of meandering fluvial architecture: Examples from the Bartlesville Sandstone, Oklahoma. Abstract: *AAPG Conference, San Antonio,* American Association of Petroleum Geologists.

Ketter, F.J. (1991). The Esmond Forbes and Gordon Fields, Blocks 43/8a, 43/13a, 43/15a, 43/20a, UK North Sea. In: *United Kingdom Oil and Gas: 25 Year Commemorative Volume* (ed. I.L. Abbotts) Memoir 14. London: Geological Society.

Kong, X. and Ohadi, M. (2010). Applications of micro and nano technologies in the oil and gas industry: Overview of the recent progress. Abu Dhabi International Petroleum Exhibition and Conference, 1-4 November, Abu Dhabi, UAE. Society of Petroleum Engineers.

Kulpecz, A.A. and van Geuns, L.C. (1988). Forties Field (UK): Geological modelling of a turbidite sequence for reservoir simulation. *AAPG Bulletin* 72 (2): 209. American Association of Petroleum Geologists.

Kuykendall, M.D. and Matson, T.E. (1992). Glenn Pool Field – USA, Northeast Oklahoma Platform, Oklahoma. In: *Stratigraphic Traps III (Treatise of Petroleum Geology Atlas of Oil and Gas Fields)*, vol. 1992 (ed. N. Foster), 155–188. Tulsa, OK: American Association of Petroleum Geologists.

Langley, G. (1987). The Forties Field development and production strategy, Proceedings of Reservoir Management in Field Development and Production Conference, Stavanger: Norwegian Petroleum Society.

Leach, H.M., Herbert, N., Los, A., and Smith, R.L. (1999). The Schiehallion development. In: Fleet, A.J. and Boldy, S.A.R. (eds) *Petroleum Geology of Northwest Europe: Proceedings of the 5th Conference* 683–692. Petroleum Geology '86 Ltd., Geological Society, London.

Leonard, A., Jolley, E., Carter, A. et al. (2000). Lessons learned from the management of basin floor fan reservoirs in the UKCS. In: *GCSSEPM Foundation 20th Annual Research Conference – Deep-Water Reservoirs of the World, December 3-6, 2000,* 478–501

Leverett, M.C. (1941). Capillary behaviour in porous solids. *Transactions of the AIME* 142: 159–172.

Long, A., Bingwen, D. and Kajl, B. (2002). PESA – Technical Focus, Petroleum Exploration Society of Australia.

Lorenz, P.B., Trantham, J.C., and Zornes, D.R. (1984). A postflood evaluation of the North Burbank surfactant/polymer pilot. Fourth Symposium on Enhanced Oil Recovery, Tulsa, p. 197-209: SPE Paper 12695.

Noel, P. and Taylor, N. (2018). Beyond Laggan–Tormore: Maximising economic recovery from gas infrastructure West of Shetland. In: *Petroleum Geology of NW Europe: 59 Years of Learning – Proceedings of the 8th Petroleum Geology Conference,* 455–464. Petroleum Geology Conferences. Geological Society, London.

Oil & Gas Journal (2013). http://www.ogj.com/articles/print/volume-111/issue-7a/general-interest/carbon-dioxide-injection-starts-in-oklahoma-s.html.

Ringrose, P.S. (2008). Total property modelling: Dispelling the net-to-gross myth. *SPE Reservoir Evaluation and Engineering* 11: 866–873.

Ringrose, P. and Bentley, M. (2015). *Reservoir Model Design: A Practitioner's Guide.* Dordrecht: Springer Science + Business Media BV.

Rose, P., Barker, G., Koster, K., and Pyle, J. (2011). Forties infill drilling eight years on: Continued success through the application of thorough development geoscience driven by 4D seismic. *SPE 145433* – Society of Petroleum Engineers.

Satter, A. and Thakur, G. (1994). *Integrated Petroleum Reservoir Management: A Team Approach.* Tulsa, OK: PennWell Publishing Company.

Seccombe, J., Lager, A., Jerauld, G. et al. (2010). Demonstration of low-salinity EOR at interwell scale, Endicott Field, Alaska. *SPE 129692* – Society of Petroleum Engineers.

Shepard, M. (1991). The Magnus Field, Block 211/7a, 12a, UK North Sea. In: *United Kingdom Oil and Gas: 25 Year Commemorative Volume* (ed. I.L. Abbotts) Memoir 14. London: Geological Society.

Simmons, M. (2016). The Wessex Basin: Petroleum geology 101 in the field. Neftex Exploration Insights Magazine. 17–23

Smith, L. and Perras, L. (1998). The Brimmond reservoir extending the Forties Field into the millennium by maximising profits without expensive rig upgrades. SPE 50383 In: *Proceedings SPE International conference on Horizontal Well Technology*, Calgary. Society of Petroleum Engineers.

SPE (PRMS) (2018). *Petroleum Reserves Management System*. Society of Petroleum Engineers www.spe.org.

Steele, R.P., Allan, R.M., Allinson, G.J., and Booth, A.J. (1993). Hyde: A proposed field development in the southern North Sea using horizontal wells. In: *Petroleum Geology of Northwestern Europe: Proceeding of the 4th Conference* (ed. J.R. Parker), 1465–1472. London: Geological Society.

Vaughan, O., Jones, R., and Plahn S. (2007). Reservoir management aspects of the rejuvenation of the Forties Field, UKCS. *SPE 109012* – Society of Petroleum Engineers.

Wahidiyat, E. (2010). Update on ESP operation at BP Wytch Farm oilfield. European Artificial Lift Forum, 17–18 February, 18pp.

Weber, K.J. (1986). How heterogeneity affects oil recovery. In: *Reservoir Characterization* (eds. L.W. Lake and H.B.J. Carroll), 487–544. Orlando, FL: Academy Press.

Weber, K.J. and van Geuns, L.C. (1990). Framework for constructing clastic reservoir simulation models. *Journal of Petroleum Technology* 42: 1248–1297.

Wehr, F. and Brasher, L.D. (1996). Impact of sequence based correlation style on reservoir model behaviour, lower Brent Group, North Cormorant Field, UK North Sea Graben. In: *High Resolution Sequence Stratigraphy: Innovations and Applications*, vol. 104 (eds. J.A. Howell and J.F. Aitken), 115–128. London: Geological Society of London Special Publication.

Wills, J.M. (1991). The Forties Field, Block 21/10, 22/6a, UK North Sea. In: *United Kingdom Oil and Gas Fields, 25 Years Commemorative Volume* (ed. I.L. Abbotts), 301–308. Geological Society.

Wills, J.M. and Peattie, D.K. (1990). The Forties Field and the evolution of a reservoir management strategy. In: *North Sea Oil and Gas Reservoirs - II: The Norwegian Institute of Technology*, 1–23. London: Graham and Trotman.

Worthington, P.F. (2002). Application of saturation-height functions in integrated reservoir description. In: *Geological Application of Well Logs* (eds. M. Lovell and N. Parkinson), AAPG Methods in Exploration No.13, 75–89. AAPG.

Zornes, D.R., Cornelius, A.J., and Long, H.Q. (1986). An overview and evaluation of the North Burbank Unit Block A Polymer Flood Project, Osage County, Oklahoma. *SPE 14113* – Society of Petroleum Engineers.

Bibliography

Amaefule, J.O., Altunbay, M., Tiab, D. et al. (1993). Enhanced reservoir description: using core and log data to identify hydraulic (flow) units and predict permeability in uncored intervals/wells. SPE 68th ATCE, Houston, SPE26435

Cosentino, L. (2001). *Integrated Reservoir Studies*. Paris: Editions Technip.

Gluyas, J. and Swarbrick, R. (2004). *Petroleum Geoscience*. Malden, MA: Blackwell Science Ltd.

Mirault, R. and Dean, L. (2007–2008). *Reservoir Engineering for Geologists*. Calgary: Canadian Society of Petroleum Geologists.

Rose, P.R. (2001). *Risk Analysis and Management of Petroleum Exploration Ventures*. AAPG Methods in Exploration Series, No. 12. Tulsa, OK: American Association of Petroleum Geologists.

Seba, R.D. (1998). *Economics of Worldwide Petroleum Production*. Tulsa, OK: OGCI Inc.

Xiangling, K. and Ohadi, M.M. (2010). Applications of Micro and Nano Technologies in the Oil and Gas Industry – Overview of the Recent Progress. *Society of Petroleum Engineers* - 14th Abu Dhabi International Petroleum Exhibition and Conference 2010, ADIPEC 2010.

Index

Reservoir Management: A Practical Guide, First Edition. Steve Cannon.
© 2021 John Wiley & Sons Ltd. Published 2021 by John Wiley & Sons Ltd.